Wireless Networks

Series Editor

Xuemin Sherman Shen, University of Waterloo, Waterloo, ON, Canada

The purpose of Springer's Wireless Networks book series is to establish the state of the art and set the course for future research and development in wireless communication networks. The scope of this series includes not only all aspects of wireless networks (including cellular networks, WiFi, sensor networks, and vehicular networks), but related areas such as cloud computing and big data. The series serves as a central source of references for wireless networks research and development. It aims to publish thorough and cohesive overviews on specific topics in wireless networks, as well as works that are larger in scope than survey articles and that contain more detailed background information. The series also provides coverage of advanced and timely topics worthy of monographs, contributed volumes, textbooks and handbooks.

Yuanwei Liu • Xidong Mu • Jiaqi Xu •
Junshan Zhang

Simultaneously Transmitting and Reflecting Surfaces for Wireless Communications

Springer

Yuanwei Liu
Department of Electrical and Electronic Engineering
The University of Hong Kong
Pokfulam, Hong Kong

Xidong Mu
School of Electronics, Electrical Engineering and Computer Science
Queen's University Belfast
Belfast, UK

Jiaqi Xu
Henry Samueli School of Engineering
University of California, Irvine
Irvine, CA, USA

Junshan Zhang
Electrical and Computer Engineering Department
University of California, Davis
Davis, CA, USA

ISSN 2366-1186 ISSN 2366-1445 (electronic)
Wireless Networks
ISBN 978-3-031-76847-7 ISBN 978-3-031-76848-4 (eBook)
https://doi.org/10.1007/978-3-031-76848-4

This Springer imprint is published by the registered company Springer Nature Switzerland AG
The registered company address is: Gewerbestrasse 11, 6330 Cham, Switzerland

STARS: An idea of 360° smart radio environment for realizing sustainable, ubiquitous, and green future wireless communications.

Preface

Simultaneously transmitting and reflecting surfaces (STARS) are emerging as one promising technology among the family of reconfigurable intelligent surfaces (RISs). Compared to conventional transmissive/reflective RISs, STARS integrates the transmission and reflection functionalities into one single RIS and realize the novel 360° smart radio environment. As a result, STARS can improve the communication performance of users located on different sides. In this book, we provide a comprehensive introduction of STARS and the employment of STARS in wireless communications. This book consists of three parts. In Part I, we discuss the fundamentals of STARS, including signal models, channel models, different categories, and hardware implementations. In Part II, we discuss about STARS-aided wireless communications from the perspective of performance analysis, beamforming design, machine learning empowered STARS, and STARS-aided integrated sensing and communications. In Part III, we highlight the future research directions and challenges of STARS. We hope that this book can attract more research efforts to unlock the full potential of STARS in 6G and beyond.

Pokfulam, Hong Kong — Yuanwei Liu
Belfast, UK — Xidong Mu
Irvine, CA, USA — Jiaqi Xu
Davis, CA, USA — Junshan Zhang
June 2024

Acknowledgments

We would like to express our sincere gratitude to the Series Editor, Prof. Xuemin (Sherman) Shen, and all the editorial staff at Springer for producing this book.

Pokfulam, Hong Kong — Yuanwei Liu
Belfast, UK — Xidong Mu
Irvine, CA, USA — Jiaqi Xu
Davis, CA, USA — Junshan Zhang
June 2024

Contents

Acronyms

2D	Two-dimensional
3D	Three-dimensional
6G	Sixth-generation
AFFL	Access-Free Federated Learning
AI	Artificial Intelligence
AL	Augmented Lagrangian
AoA	Angle-of-Arrival
AoD	Angle-of-Departure
AP	Access Point
AWGN	Additive White Gaussian Noise
BCD	Block Coordinate Descent
BER	Bit Error Rate
BS	Base Station
CRB	Cramér–Rao Bound
C&S	Communication and Sensing
CSI	Channel State Information
DDPG	Deep Deterministic Policy Gradient
DNN	Deep Neural Network
DOA	Direction of Arrival
DoFs	Degrees-of-Freedom
DPPO	Distributed Proximal Policy Optimization
DQN	Deep-Q Network
DRL	Deep Reinforcement Learning
EE	Energy Efficiency
EM	Electromagnetic
ES	Energy Splitting
FDMA	Frequency-Division Multiple Access
FL	Federated Learning
FPGA	Field-Programmable Gate Array
GSTC	Generalized Sheet Transition Conditions
i.i.d.	identically distributed

IoT	Internet-of-Things
ISAC	Integrated Sensing And Communications
KKT	Karush-Kuhn-Tucker
LoS	Line-of-Sight
MDP	Markov Decision Process
MEC	Mobile Edge Computing
ML	Machine Learning
mmWave	millimeter Wave
MS	Mode Switching
MSE	Mean Square Error
NFC	Near-Field Communications
NLoS	Non-Line-of-Sight
NOMA	Non-Orthogonal Multiple Access
OMA	Orthogonal Multiple Access
PDD	Penalty Dual Decomposition
PDF	Probability Density Function
PIN	Positive Intrinsic Negative
PLS	Physical Layer Security
PPO	Proximal Policy Optimization
PSC	Phase-Shift Configuration
QoS	Quality-of-Service
RC	Reflection Coefficient
RF	Radio Frequency
RIS	Reconfigurable Intelligent Surface
RL	Reinforcement Learning
Rx	Receiver
SC	Superposition Coding
SCA	Successive Convex Approximation
SDP	Semidefinite Program
SDR	Semidefinite Relaxation
SIC	Successive Interference Cancellation
SINR	Signal-to-Interference-plus-Noise Ratio
SNR	Signal-to-Noise Ratio
SRE	Smart Radio Environment
SWIPT	Simultaneous Wireless Information and Power Transfer
STARS	Simultaneously Transmitting And Reflecting Surface
TC	Transmission Coefficient
THz	Terahertz
TL	Transfer Learning
TRC	Transmission and Reflection Coefficient
TS	Time Switching
Tx	Transmitter
ULA	Uniform Linear Array
UPA	Uniform Planar Array
WMMSE	Weighted Minimum Mean Square Error

Chapter 1
Introduction

Abstract In this chapter, we will discuss the background of reconfigurable intelligent surfaces (RISs) and introduce the novel idea of simultaneously transmitting and reflecting surfaces (STARSs) in the family of RISs.

1.1 Background of RISs

With the rapid development of metasurfaces and the corresponding fabrication technologies, reconfigurable intelligent surfaces (RISs) and their diverse variants (Di Renzo et al., 2020; Liu et al., 2021; Huang et al., 2020a) have emerged as promising techniques for sixth-generation (6G) wireless networks. Generally speaking, RISs are two-dimensional (2D) structures and are comprised of a large number of low-cost reconfigurable elements. By employing a smart controller (e.g., a field-programmable gate array (FPGA)) attached to the RIS, both the phase and even the amplitude of these reconfigurable elements can be beneficially controlled, thus reconfiguring the propagation of the incident wireless signals and realizing a "Smart Radio Environment (SRE)" (Di Renzo et al., 2020). Since no radio frequency (RF) chains are required, RISs are more economical and environmentally friendly than the family of conventional multi-antenna and relaying technologies. Given these beneficial RIS characteristics, extensive industrial and academic research efforts have been devoted to the investigation of RISs, including but not limited to the design of energy-efficient communication (Huang et al., 2019), the support of non-orthogonal multiple access (NOMA) communications (Mu et al., 2020), the mitigation of blockages in millimeter wave (mmWave)/terahertz (THz) communications (Akyildiz et al., 2018), and artificial intelligence (AI) aided implementations (Huang et al., 2020b).

Although some recent studies have considered both transmissive and reflective metasurfaces for wireless communications (as will be discussed below), the existing studies mainly focus on RISs whose only function is to reflect an incident signal, indicating that both the source and the destination have to be at the same side of the RISs (Huang et al., 2019; Mu et al., 2020; Akyildiz et al., 2018; Huang et al., 2020b), i.e., within the same *half-space* of the SRE. This topological constraint limits the

Y. Liu et al., *Simultaneously Transmitting and Reflecting Surfaces for Wireless Communications*, Wireless Networks, https://doi.org/10.1007/978-3-031-76848-4_1

flexibility of employing conventional RISs, and to address this issue, the concept of simultaneously transmitting and reflecting surface (STARS)[1] was proposed (Xu et al., 2021), where the incident wireless signals can be reflected within the half-space of the SRE at the same side of the RIS, but they can also be transmitted into the other side of the RIS. As a result, a *full-space* SRE can be created by STARSs.

1.2 From Conventional Reflecting-Only RISs to STARSs

We commence by providing a brief introduction to three types of signal propagation achieved by metasurfaces, namely full reflection, full transmission, as well as simultaneous transmission and reflection, as shown in Fig. 1.1a–c. For the full reflection scenario of Fig. 1.1a, the incident signals are completely reflected and cannot penetrate the surface. This type of wireless signal manipulation is widely investigated for conventional reflecting-only RISs. By contrast, for the full transmission scenario of Fig. 1.1b, all incident signals pass through the surface into the transmission space, while no signal is reflected. Finally, for the simultaneous transmission and reflection scenario of Fig. 1.1c, the incident signals are divided into two parts by the surface. Part of the signal is reflected in the reflection space, while the remaining part is radiated into the transmission space, thus facilitating the full-space manipulation of signal propagation. In particular, STARSs were developed to achieve simultaneous transmission and reflection functions and thus provide full-space coverage.

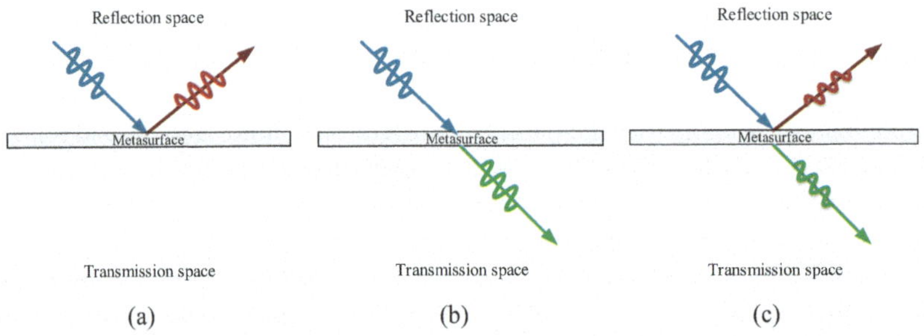

Fig. 1.1 Illustration of signal propagation types of metasurfaces. (**a**) Full reflection (conventional reflecting-only RIS). (**b**) Full transmission. (**c**) Simultaneous transmission and reflection

[1] From a physical perspective, transmission and reflection refer to the two fundamental physical processes of electromagnetic (EM) wave propagation through the surface. In this book, we use the term "transmission" to describe the passive signal refraction process instead of the process of actively producing a signal.

Considering the above unique features, the employment of STARSs has the following advantages in wireless communication systems: (1) Thanks to their capability of simultaneously transmitting and reflecting the incident signals, the coverage of STARSs is extended to the entire space, thus serving both half-spaces using a single RIS, which is not possible for conventional reflecting-only RISs. (2) STARSs provide enhanced degrees of freedom (DoFs) for signal propagation manipulation, which significantly increases the design flexibility in satisfying stringent communication requirements. (3) Since STARSs can be designed to be optically transparent (NTT DOCOMO, 2020), they are aesthetically pleasing and readily compatible with existing building structures, such as windows. Therefore, STARSs will have no undesired aesthetic effect, which is of vital importance for practical implementations.

Against the above observations, STARSs are one promising technology for 6G among the family of RISs. Before reaping the benefits of STARSs in future wireless networks, a comprehensive discussion of STARSs is essential to provide an in-depth understanding of the fundamentals, performance analysis, and beamforming design of STARSs as well as an inspiring discussion of possible applications of STARSs. This provides the main motivation for writing this book.

References

Akyildiz, I. F., Han, C., & Nie, S. (2018). Combating the distance problem in the millimeter wave and terahertz frequency bands. *IEEE Communications Magazine, 56*, 102–108.

Di Renzo, M., Zappone, A., Debbah, M., Alouini, M.-S., Yuen, C., de Rosny, J., & Tretyakov, S. (2020). Smart radio environments empowered by reconfigurable intelligent surfaces: How it works, state of research, and the road ahead. *IEEE Journal on Selected Areas in Communications, 38*, 2450–2525.

Huang, C., Hu, S., Alexandropoulos, G. C., Zappone, A., Yuen, C., Zhang, R., Renzo, M. D., & Debbah, M. (2020a). Holographic MIMO surfaces for 6G wireless networks: Opportunities, challenges, and trends. *IEEE Wireless Communications, 27*, 118–125.

Huang, C., Mo, R., & Yuen, C. (2020b). Reconfigurable intelligent surface assisted multiuser MISO systems exploiting deep reinforcement learning. *IEEE Journal on Selected Areas in Communications, 38*, 1839–1850.

Huang, C., Zappone, A., Alexandropoulos, G. C., Debbah, M., & Yuen, C. (2019). Reconfigurable intelligent surfaces for energy efficiency in wireless communication. *IEEE Transactions on Wireless Communications, 18*, 4157–4170.

Liu, Y., Liu, X., Mu, X., Hou, T., Xu, J., Di Renzo, M., & Al-Dhahir, N. (2021). Reconfigurable intelligent surfaces: Principles and opportunities. *IEEE Communications Surveys and Tutorials, 23*, 1546–1577.

Mu, X., Liu, Y., Guo, L., Lin, J., & Al-Dhahir, N. (2020). Exploiting intelligent reflecting surfaces in NOMA networks: Joint beamforming optimization. *IEEE Transactions on Wireless Communications, 19*, 6884–6898.

NTT DOCOMO. (2020). DOCOMO conducts world's first successful trial of transparent dynamic metasurface. https://www.nttdocomo.co.jp/english/info/media_center/pr/2020/0117_00.html

Xu, J., Liu, Y., Mu, X., & Dobre, O. A. (2021). STAR-RISs: Simultaneous transmitting and reflecting reconfigurable intelligent surfaces. *IEEE Communications Letters, 25*, 3134–3138.

Part I
Fundamentals of STARS

Chapter 2
Signal and Channel Models of STARS

Abstract In this chapter, we will introduce the basic signal model and channel models for employing STARSs in wireless communications.

2.1 Basic Signal Modeling of STARS

2.1.1 Independent Phase-Shift Model

As shown in Fig. 2.1, a schematic representation of the structure of STARSs is presented. Based on the field equivalence principle (Liu et al., 2021), as the STARS elements are excited by the incident signal, the transmitted and reflected signals can be equivalently treated as waves radiated from the time-varying surface equivalent electric currents $\mathbf{J}_s$ and magnetic currents $\mathbf{K}_s$ (also referred to as the bound currents). Within each element, the strengths and distribution of these surface equivalent currents are determined by the incident narrowband signal s_m as well as the local surface averaged electric and magnetic impedances Y_m and Z_m. Assume that the STARS produces both transmitted and reflected signals with the same polarization. At the mth element, these signals can be expressed as

$$s_m^T = T_m s_m, \; s_m^R = R_m s_m, \tag{2.1}$$

where T_m and R_m are the transmission and reflection coefficients (TRCs) of the mth element, respectively. According to the law of energy conservation, for passive STARS elements, the following constraint on the local TRCs must be satisfied:

$$|T_m|^2 + |R_m|^2 \leq 1. \tag{2.2}$$

According to EM theory, the phase delays of both the transmitted and reflected fields are related to Y_m and Z_m. In Fig. 2.1, the reconfigurability of the element is reflected in the change of the surface impedances, since the TRCs of the mth element are related to the surface impedances as $T_m = \frac{2-\eta_0 Y_m}{2+\eta_0 Y_m} - R_m$, and

Y. Liu et al., *Simultaneously Transmitting and Reflecting Surfaces for Wireless Communications*, Wireless Networks, https://doi.org/10.1007/978-3-031-76848-4_2

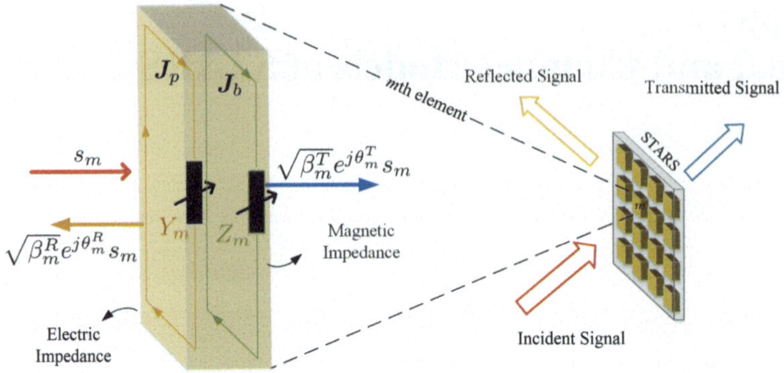

Fig. 2.1 Schematic illustration of the STARS

$R_m = -\frac{2(\eta_0^2 Y_m - Z_m)}{(2+\eta_0 Y_m)(2\eta_0+Z_m)}$, where η_0 is the impedance of free space (Estakhri & Alu, 2016). From the perspective of metasurface design, supporting the magnetic currents is the key to achieving independent control of both the transmitted and reflected signals. According to La Spada et al. (2019), single-layered RISs with nonmagnetic elements can only produce identical radiation on different sides, which is referred to as the symmetry limitation. By introducing the equivalent surface electric and magnetic currents into the model, the proposed hardware model is able to independently characterize the transmission and reflection of each element. To facilitate the design of STARSs in wireless communication systems, we rewrite these narrowband frequency-flat coefficients in the form of their amplitudes and phase shifts as follows:

$$T_m = \sqrt{\beta_m^T} e^{j\theta_m^T}, \ R_m = \sqrt{\beta_m^R} e^{j\theta_m^R}, \tag{2.3}$$

where $\beta_m^T, \beta_m^R \in [0, 1]$ are real-valued amplitude coefficients satisfying $\beta_m^T + \beta_m^R \leq 1$, and $\theta_m^T, \theta_m^R \in [0, 2\pi)$ are the phase shifts introduced by element m for the transmitted and reflected signals, $\forall m \in \{1, 2, \cdots, M\}$. For the independent phase-shift model, different from the amplitude coefficients (β_m^T and β_m^R) that are coupled due to the law of energy conservation, we assume that the phase-shift coefficients (i.e., θ_m^T and θ_m^R) can be independently adjusted.

2.1.2 *Coupled Phase-Shift Model*

In this subsection, we continue to introduce the coupled phase-shift model for STARSs. Note that the independent phase-shift model presented in the previous subsection imposes no constraints on the surface impedance Y_m and Z_m. However, for passive lossless STARS elements, these impedances need to be purely imaginary,

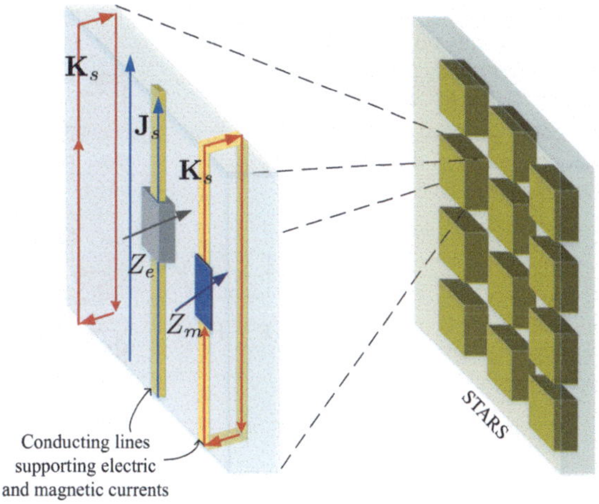

Fig. 2.2 Conceptual illustration of the hardware model of the STARS

which gives rise to the coupled phase-shift model. Before presenting the coupled phase-shift model, we first introduce some prerequisites from physics to capture the characteristics of STARSs.

Consider an STARS whose elements are made of a single-layered passive material. Upon being illuminated by the incident EM wave, multiple currents are induced in each element, cf. Fig. 2.2. For STARS modeling, it is sufficient to consider only two types of currents,[1] namely the electric current and the magnetic current (Vágó, 1994). The electric current is generated by the flow of free electrical charges within the conducting part of the element media. Let $\mathbf{J}_s$ denote the electric current density. As shown in Fig. 2.3, since the electric field component of the impinging EM field is oscillating, $\mathbf{J}_s$ also oscillates and thus produces EM radiation with electric field $\mathbf{E}^J$, and magnetic fields $\mathbf{H}_1^J$ and $\mathbf{H}_2^J$ on the two sides of the STARS, respectively. The magnetic current density, $\mathbf{K}_s$, is generated by vortex (circular) currents within each STARS element (Rothwell & Cloud, 2018). These currents are induced by the oscillating magnetic component of the EM field, and they produce EM radiation with electric fields $\mathbf{E}_1^K$ and $\mathbf{E}_2^K$, and magnetic field $\mathbf{H}^K$. The strengths of these two types of currents are assumed to be proportional to the sum of the incident and radiated fields at the STARS. Thus, the density of the electric and magnetic currents are respectively given by

$$\mathbf{J}_s = Y_m(\mathbf{E}^{inc} + \mathbf{E}^J), \tag{2.4}$$

$$\mathbf{K}_s = Z_m(\mathbf{H}^{inc} + \mathbf{H}^K), \tag{2.5}$$

where Y_m and Z_m are the scalar electric and magnetic impedances of a particular element, respectively, and $\mathbf{E}^{inc}$ and $\mathbf{H}^{inc}$ are the electric and magnetic components

[1] This is because, in terms of radiation, the electric/magnetic charge layers and electric/magnetic current layers are equivalent.

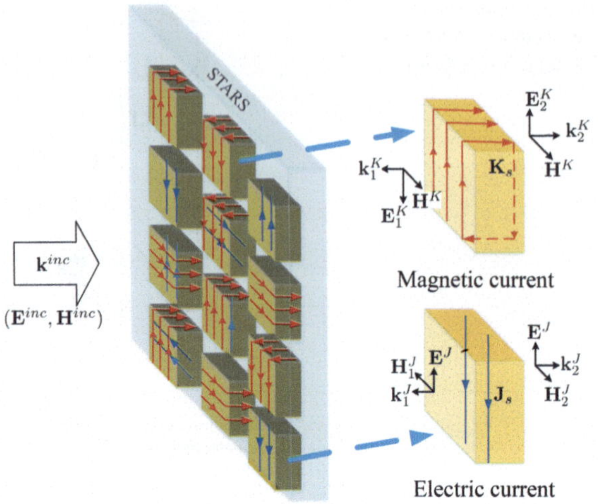

Fig. 2.3 EM radiations of STARS elements carrying induced currents $\mathbf{J}_s$ and $\mathbf{K}_s$, where $\mathbf{k}^{inc}$, $\mathbf{k}_1^J$, $\mathbf{k}_2^J$, $\mathbf{k}_1^K$, and $\mathbf{k}_2^K$ denote the wave vectors (wave numbers) of the corresponding waves

of the incident field, respectively. To connect the EM fields on both sides of the STARS, we take into account the boundary conditions of the EM field as follows Vágó (1994):

$$\mathbf{n} \times (\mathbf{H}_1^J - \mathbf{H}_2^J) = \mathbf{J}_s \text{ and } \mathbf{n} \times (\mathbf{E}_1^K - \mathbf{E}_2^K) = \mathbf{K}_s, \tag{2.6}$$

where $\mathbf{n}$ is the unit vector perpendicular to the STARS.

The strengths of the transmitted and reflected signals are determined by the magnitudes of the electric components of the transmitted and reflected EM fields. Assuming a vertically polarized wireless signal, the $\mathbf{E}$ fields are in the x-direction and the $\mathbf{H}$ fields are in the y-direction (see Fig. 2.3). Omitting the superscripts, we can rewrite these fields as $\mathbf{E} = \hat{\mathbf{x}}E$ and $\mathbf{H} = \hat{\mathbf{y}}H$, where $\hat{\mathbf{x}}$ and $\hat{\mathbf{y}}$ are the unit vectors in x-direction and y-direction, respectively, and E and H denote the complex amplitudes of the corresponding fields. Specifically, considering an element with both induced electric and magnetic currents ($\mathbf{J}_s$ and $\mathbf{K}_s$), the TRCs can be expressed as follows:

$$T = \beta^T \cdot e^{j\theta^T} = (E^J + E_2^K + E^{inc})/E^{inc}, \tag{2.7}$$

$$R = \beta^R \cdot e^{j\theta^R} = (E^J + E_1^K)/E^{inc}, \tag{2.8}$$

where β^T and β^R are the real-valued transmission and reflection amplitudes, respectively, and θ^T and θ^R are the corresponding phase-shift values for transmission and reflection.

Next, we study the correlation between θ^T and θ^R for a lossless STARS element. According to EM theory, the change in energy of an EM field within an arbitrary volume V is given as follows Rothwell and Cloud (2018):

$$\frac{\mathrm{d}W}{\mathrm{d}t} = -\int_{(\Sigma)} (\mathbf{E} \times \mathbf{H})\, \mathrm{d}\Sigma - \int_{(V)} \mathbf{J} \cdot \mathbf{E}\, \mathrm{d}V, \tag{2.9}$$

where W is the EM energy within the volume V, d/dt denotes the time derivative, $\mathbf{E}$ and $\mathbf{H}$ denote the EM fields at the boundary of V, $\mathbf{J}$ denotes the surface electric and magnetic current densities, operators $\times$ and $\cdot$ denote the cross product and dot product between two vectors, respectively, and Σ denotes the closed surface of the chosen volume V. The second term in (2.9) is the integration of the Poynting vector evaluated at the closed surface Σ, which reflects the energy loss due to EM radiation. The third term is volume integration, and it reflects the Ohmic heating and the power of other non-electrostatic forces within volume V. According to (2.9), if we choose V as the volume of an STARS element, the second term represents the EM energy radiated toward the target users, while the third term represents the energy loss in the volume. Thus, a locally lossless element requires that the time-averaged value of $\mathbf{J} \cdot \mathbf{E}$ is equal to zero for the entire element. In the context of (2.4) and (2.5), this indicates that Y_m and Z_m should have purely imaginary values. By exploiting (2.4)–(2.9), we obtain the coupled phase-shift model for the passive lossless STARS as given in the following proposition.

Proposition 2.1 (Coupled Phase-Shift Model) *For the passive lossless STARS, the transmission and reflection amplitudes, β_m^T and β_m^R, and the phase shifts, θ_m^T and θ_m^R, of the mth element have to meet the following constraints:*

$$\beta_m^T = 1 - \beta_m^R, \tag{2.10}$$

$$\theta_m^R - \theta_m^T = \frac{\pi}{2} + \nu_m \pi, \ \nu_m = 0 \ or \ 1, \ \forall m = 1, 2, \cdots, M, \tag{2.11}$$

where M denotes the total number of STARS elements. ν_m is referred to as the auxiliary bit[2] *for the lossless STARS in the following and provides an additional degree of freedom that links the possible phase-shift values between the TRCs.*

Proof The relation in (2.10) is a direct consequence of the law of energy conservation, i.e., the sum of the radiation power of both sides should be equal to the power of the incident signal (Zhu et al., 2014). For (2.11), by exploiting (2.4)–(2.8), it can be shown that the complex-valued TRCs of the passive STARS elements have to satisfy $|T + R| = 1$ or $|T - R| = 1$, or equivalently, $\sqrt{\beta_m^R}\sqrt{1 - \beta_m^R}\cos(\theta_m^R - \theta_m^T) = 0$. Thus, for nonzero β_m^R, the phase-shift difference must fulfill

$$|\theta_m^R - \theta_m^T| = \pi/2 \text{ or } 3\pi/2. \tag{2.12}$$

□

[2] ν_m can only take on two values, 0 or 1. Thus, its value represents the two possible phase differences between θ_m^R and θ_m^T, i.e., the phase difference is either $\pi/2$ or $3\pi/2$.

Remark 2.1 The coupled phase-shift model in (2.12) is a consequence of Maxwell's Equations and the boundary conditions. We note that the independent phase-shift model is achievable if the impedances Y_m and Z_m in (2.4) and (2.5) can take on arbitrary complex values. However, a nonzero real part of Y_m and Z_m indicates that the STARS requires active or lossy elements, which might significantly increase the manufacturing cost.

2.1.3 Channel Models

In this section, we will introduce the communication channels of STARSs. As shown in Fig. 2.4, a narrowband STARS-aided wireless communication system is considered, where the transmitter (Tx) and receivers (Rxs) are equipped with a single antenna and the STARS consists of M reconfigurable elements. As an initial analysis to demonstrate the extended coverage of the STARS, we consider a two-user setup. In particular, the Tx sends information to the two Rxs employing orthogonal frequency bands of equal sizes. Suppose that Rx T is located on the opposite side of Tx, with respect to the STARS, and thus, it can only receive the transmitted signals of the STARS. Rx R and Tx are located on the same side of STARS. Let h^{χ} denote the direct link between Tx and Rx T or Rx R, where the notation $\chi \in \{T, R\}$ is an indicator representing "T" or "R." Let g^{χ} denote the effective channel between Tx and Rx χ through STARS transmission or reflection.

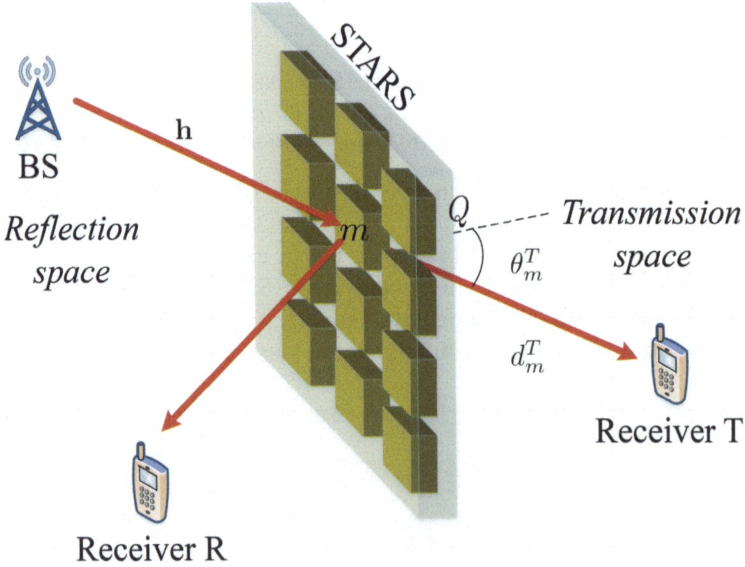

Fig. 2.4 Illustration of the considered STARS-aided wireless communication system

2.1.3.1 Far-Field Channel Model

In this case, the Rxs are located in the far-field region of the STARS. We denote the channel between Tx and the STARS by $\mathbf{h} = (h_1, \cdots, h_M)^T$, where h_m is the channel between Tx and the mth element. In addition, let $\mathbf{r}^\chi = (r_1^\chi, \cdots, r_M^\chi)^T$ denote the channel between STARS and Rx χ. Since both Rxs are located in the far-field region of the STARS, the ray-tracing technique can be adopted by studying a number of M geometrical rays, each corresponding to a multipath signal propagating through an element. This leads to the channel model as follows:

$$g^\chi = (\mathbf{r}^\chi)^H \text{diag}(\sqrt{\beta_1^\chi} e^{j\theta_1^\chi}, \cdots, \sqrt{\beta_M^\chi} e^{j\theta_M^\chi})\mathbf{h}, \tag{2.13}$$

where $\mathbf{a}^H$ denotes the conjugate transpose of a complex vector $\mathbf{a}$. For convenience, we denote $\text{diag}(T_1, T_2, \cdots, T_M)$ by $\boldsymbol{\Theta}^T$ and $\text{diag}(R_1, R_2, \cdots, R_M)$ by $\boldsymbol{\Theta}^R$. In addition, in the far-field channel model, $\mathbf{r}^\chi$ and $\mathbf{h}$ can be written in the form of path loss (large-scale fading) multiplied with the normalized small-scale fading channels. The large-scale fading depends on the distance between the Tx, the STARS, and Rxs, while the small-scale fading depends on the scattering environment. Let d^χ denote the distances between the STARS and Rx χ, and d_0 denote the distance between Tx and the STARS. The channel gains can be expressed as

$$|g^\chi| = \frac{1}{(d^\chi)^{\alpha_\chi}(d_0)^{\alpha_0}} |(\widetilde{\mathbf{r}}^\chi)^H \boldsymbol{\Theta}^\chi \widetilde{\mathbf{h}}|, \tag{2.14}$$

where $\widetilde{\mathbf{r}}^\chi$ and $\widetilde{\mathbf{h}}$ are the corresponding small-scale fading components. In addition, α_χ and α_0 are the path loss coefficients of the corresponding channels.

2.1.3.2 Near-Field Channel Model

In scenarios where Rxs are located within the near-field region of the STARS, the conventional ray-tracing technique-based channel models cannot be adopted. Based on the Huygens-Fresnel principle (Liu et al., 2021), A. Fresnel and Kirchhoff arrived at the analytical result that is known as the Fresnel-Kirchhoff diffraction formula. As illustrated in Fig. 2.4, the EM signal at the Rx can be calculated by summing up the contribution of each element on the wavefront. The wavefront is chosen as the plane where the STARS is located.

$$g^\chi = \frac{1}{j\lambda} \iint_{(\Sigma)} U^\chi(Q) F(\phi^\chi) \frac{e^{2j\pi d_m^\chi/\lambda}}{d^\chi} d\Sigma, \tag{2.15}$$

where j is the imaginary unit, (Σ) denotes a closed surface (wavefront) that contains the RIS elements and the scatterers in the environment, $U^\chi(Q)$ is the aperture distributions for the transmitted or reflected waves at point Q on (Σ), $F(\phi^{\chi})$ is the leaning factor at point Q, d^{χ} are the distances between Q and the Rx χ, and λ is the free-space wavelength of the signal. The integral in (2.15) can be evaluated element by element. Assuming that the aperture distribution $U^{\chi}(Q)$ is uniform within each element, at the mth element, we have $U^{T}(Q) = \theta_m^T h_m$ and $U^{R}(Q) = \theta_m^R h_m$. Thus, (2.15) can be expressed as

$$g^{\chi} = \frac{A_e}{j\lambda}\sum_m \theta_m^{\chi} h_m F(\phi_m^{\chi}) \frac{e^{2j\pi d_m^{\chi}/\lambda}}{d_m^{\chi}}, \tag{2.16}$$

where A_e is the area of each element, and ϕ_m^{χ} is the direction of Rx χ, with respect to the normal direction of STARS, as illustrated in Fig. 2.4. According to the Fresnel-Kirchhoff diffraction formula, the leaning factor is $F(\phi_m^{\chi}) = (1+\cos\phi_m^{\chi})/2$, which holds for both *Rx T* and *Rx R*. Thus, the channel gain can be expressed as

$$|g^{\chi}| = \frac{A_e}{\lambda}\left|\sum_m \theta_m^{\chi} h_m (1+\cos\phi_m^{\chi}) \frac{e^{2j\pi d_m^{\chi}/\lambda}}{2d_m^{\chi}}\right|. \tag{2.17}$$

By comparing these results with the channel gains of the far-field channel model, it can be noticed that the distances between STARS elements and Rxs cannot be treated as the same and be brought outside of the summation. In addition, the contribution of the leaning factor should be explicitly considered in the near-field model.

2.2 Summary

In this chapter, we introduced the basic signal models for STARS, namely independent and coupled phase-shift models. Moreover, both far-field and near-field channel models for employing STARSs in wireless communications were also discussed.

References

Estakhri, N. M., & Alu, A. (2016). Wave-front transformation with gradient metasurfaces. *Physical Review X, 6*, 041008.

La Spada, L., Spooner, C., Haq, S., & Hao, Y. (2019). Curvilinear metasurfaces for surface wave manipulation. *Scientific Reports, 9*, 1–10.

Liu, Y., Liu, X., Mu, X., Hou, T., Xu, J., Di Renzo, M., & Al-Dhahir, N. (2021). Reconfigurable intelligent surfaces: Principles and opportunities. *IEEE Communications Surveys & Tutorials, 23*, 1546–1577.

Rothwell, E. J., & Cloud, M. J. (2018). *Electromagnetics*. CRC Press.

Vágó, I. (1994). On the interface and boundary conditions of electromagnetic fields. *Periodica Polytechnica Electrical Engineering, 38*, 79–94.

Zhu, B. O., Chen, K., Jia, N., Sun, L., Zhao, J., Jiang, T., & Feng, Y. (2014). Dynamic control of electromagnetic wave propagation with the equivalent principle inspired tunable metasurface. *Scientific Reports, 4*, 1–7.

Chapter 3
Categories and Hardware Implementations of STARS

Abstract In this chapter, we will introduce promising techniques that may be used for implementing STARSs and hardware models for characterizing the tuning capability of STARSs at different levels of accuracy. We will also discuss the advantages and disadvantages of different channel models that can used for STARSs. Moreover, two STARS variants, namely the dual-sided STARS and the active STARS, will be discussed.

3.1 Hardware Implementations for STARSs

How the *STARS* concept can be implemented based on practical hardware designs is one of the most pivotal questions in the research of STARSs. To address this issue, in this section, we survey and categorize the possible implementation options of STARSs to achieve independent control of the transmitted and reflected signals.

There are various tunable surface designs that are potential candidates for realizing STARSs. In Di Renzo et al. (2020), the authors pointed out an intuitive difference between natural and artificial materials (RISs in general), namely that natural materials exhibit uniform EM properties along their tangential directions, while artificial materials exhibit either a periodic or quasi-periodic nature. In terms of the periodic structure, we can loosely classify the hardware implementations of STARSs into two categories, namely the patch-array-based implementations and the metasurface-based implementations. The patch-array-based implementations consist of periodic cells having sizes on the order of a few centimeters. Because of their relatively large sizes, each cell (patch) can be made tunable by incorporating either positive intrinsic negative (PIN) diodes or delay lines. By contrast, the metasurface-based implementations have periodic cells on the order of a few millimeters, possibly micrometers, or even molecular sizes. Hence, they require more sophisticated controls of their EM properties, such as conductivity and permittivity. Below, we provide a brief overview of a pair of patch-array-based implementations and two metasurface-based implementations. All these hardware implementations have had successful prototypes built or rely on strong theoretical evidence in support of their feasibility.

Y. Liu et al., *Simultaneously Transmitting and Reflecting Surfaces for Wireless Communications*, Wireless Networks, https://doi.org/10.1007/978-3-031-76848-4_3

3.1.1 Patch-Array-Based STARSs

3.1.1.1 PIN Diode-Empowered Implementations

For patch-array-based implementations, both the phase and amplitude response can be tuned by applying different bias voltages to the PIN diodes. In Zhang et al. (2022), the authors presented an STARS prototype relying on the PIN diode-empowered implementation. This implementation is the most popular design for both RISs and STARSs since PIN diodes are of low cost and are voltage-controlled. The drawback of this implementation is that since PIN diodes only have two states, namely, "ON" or "OFF," this implementation can only support a finite-cardinality TRC set. Moreover, for a given state of all the PIN diodes, the TRCs are coupled. As a result, the PIN diode-empowered implementation struggles to mimic independent control of both transmission and reflection unless a sufficiently high number of PIN diodes are used for each patch element.

3.1.1.2 Antenna-Empowered Implementations

The concept of phased-array antennas may be readily extended to STARSs with some minor modifications. According to Nemati et al. (2021), each antenna-empowered STARS cell actually consists of two antenna elements, which are connected by a tunable delay line (waveguide). The antenna elements facing the incident wave operate similarly to the reflecting-only RIS elements, but a certain fraction of the incident energy is transferred along the delay lines and it is re-radiated into the *transmission space*. The phase of the TC of each cell is determined by the length of the delay line. Thus, the phase shifts of both the reflected and the transmitted signals can be independently adjusted. However, the drawback of this implementation is that the delay line may impose considerable energy loss. If the desired phase shift of the transmitted signal is high, the amplitude of the TC will be reduced. Thus, the amplitude and phase of the TC are correlated in this implementation.

3.1.2 Metasurface-Based STARSs

3.1.2.1 DOCOMO's Smart Glass

A popular prototype of the metasurface-based STARSs is the *transparent dynamic metasurface* designed by researchers at NTT DOCOMO, Japan. According to NTT DOCOMO (2020), the metasurface supports the manipulation of 28 GHz (5G) radio signals. It allows dynamic control of both the signal's transmission and reflection while maintaining the transparency of the window. By adjusting the width of the dielectric material and its distance between substrates, the smart glass can be switched into different modes, such as full penetration (transmission), full reflection, and partial reflection. DOCOMO revealed that they are working on more sophisticated tuning techniques and will use them in future trials.

The main advantage of DOCOMO's smart glass is that owing to its transparency in the visible light frequency range, it can be aesthetically integrated into buildings. However, its drawback is that it does not have the ability to dynamically reconfigure itself as the PIN diode-empowered implementations. Moreover, adjusting the distance between substrates may affect the RC of the entire surface instead of only reconfiguring a particular element.

3.1.2.2 Graphene-Empowered Implementations

It has been widely exploited that a single graphene layer has extraordinary properties, including a beneficial EM wave response that may also be used for building STARSs. More importantly, to realize futuristic envisions such as smart surfaces-assisted visible light wireless transmission and wearable skin-like smart surfaces, we might have to rely on graphene-empowered technologies. Indeed, there are already experimental graphene-based RF devices (Zhu et al., 2013). To achieve reconfigurability, a single layer of graphene has tunable TRCs by adjusting its conductivity. Moreover, a periodic stack of graphene layers is capable of acting as a tunable spectrally selective mirror. We may summarize that for graphene-empowered STARSs, even the separation of a combined signal might become feasible based on the different carrier frequencies or polarizations. In a nutshell, graphene-empowered implementations might open extraordinary possibilities for the design of smart radio environments once their fabrication process becomes more economical.

3.1.3 STARS Implementations and Operating Frequencies

Naturally, almost all designs can only operate as desired within a certain frequency range. This is because in order for the STARS to apply the desired phase shifts and wavefront transformations to both the transmitted and reflected signals, the length of periodicity in STARSs has to *match* the wavelength of the wireless signal. For the patch-array-based implementations, the periodicity is usually chosen to be between 0.5λ and 0.7λ, where λ is the wavelength of the wireless signal (Mailloux, 2017). According to this relationship, the patch-array-based STARSs are suitable for assisting wireless communication with low carrier frequencies, e.g., sub-6 GHz. For wireless signals having higher frequencies (e.g., mmWave or THz) and for visible light communication, metasurface-based STARSs are required.

3.1.4 Summary and Outlook

In conclusion, all the above-mentioned STARS implementations achieve independent control of the transmitted and reflected signals, when an adequate number of control units are applied to each element. However, there is a trade-off between the phase/amplitude control accuracy and design complexity. Furthermore, a more accurate control can be achieved at the expense of increasing the energy consumed per adjustment and the time delay of each adjustment. In Table 3.1, we summarize the operating frequency and tuning mechanism of each STARS implementation discussed. It is worth noting that the current development of STARS prototypes is still at an early stage. For future development of the hardware design and manufacture of STARSs, we need to find solutions to make STARSs more scalable, while maintaining the tunability of each individual element. This can be achieved by leveraging the recent achievements in metasurface technology and nano-engineering (Di Renzo et al., 2020).

3.2 Hardware Models for STARSs

As discussed in Sect. 3.1, different STARS implementations are rather different in terms of their tuning mechanisms. However, we require a unified technique for modeling the effect of these surfaces on the wireless signal. Explicitly, we have to find an accurate hardware model for characterizing the EM wave response of the STARS. From a tangible physical perspective, modeling a smart surface is equivalent to the problem of studying the boundary conditions of the EM field at the surface. However, the interaction of an arbitrary field with the STARS is an intrinsically complex problem. All existing hardware models rely on certain levels of approximations based on their own different assumptions. Our next objective is to demonstrate and compare these assumptions, as well as to reveal the physical abstractions behind each model.

3.2.1 The Phase-Shift Model

The phase-shift model characterizes a smart surface using a collection of phase-shift values, or applying a specific phase shift as a function of the cell's position on the surface. This function is also often referred to as the phase profile, phase discontinuity, or phase-shift matrix, depending on the context. As illustrated in Fig. 3.1a, the physical abstractions laying the foundation of the phase-shift model are as follows: The STARS can be regarded as a periodic array of either metallic or dielectric particles. Regardless of the specific geometric and EM properties of these particles, the reflected or transmitted field radiated from the STARS can be

Table 3.1 Comparing different implementations of STARSs

Implementation	Operating frequency	STARS prototypes	Tuning mechanism	Independent transmission reflection control
Patch-array-based	Low to high frequency	PIN diode empowered	Bias voltages on PIN diodes	Difficult to achieve
		Antenna empowered	Lengths of delay lines	Can be achieved
Metasurface-based	Super high frequency to visible light frequency	DOCOMO's smart glass	Distance between substrates	Theoretically achievable
		Graphene empowered	Conductivity of graphene	Can be achieved

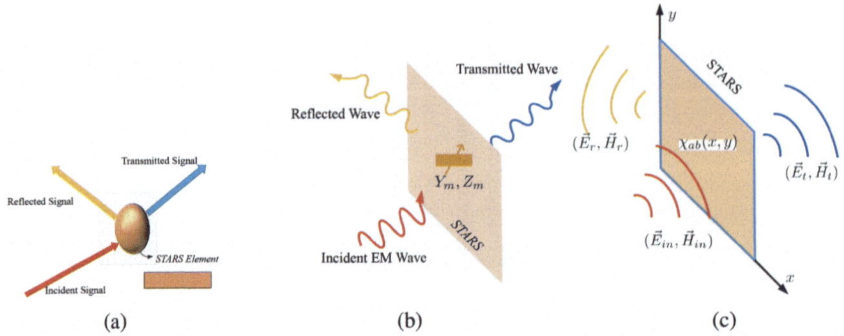

Fig. 3.1 Conceptual comparison between different models for the patch-array-based STARS implementations, where Y_m and Z_m are the surface electric and magnetic impedances, χ_{ab} is the surface susceptibility dyadics, and (**E**, **H**) are the electric and magnetic components of the corresponding EM fields. (**a**) The phase-shift model. (**b**) The load impedance model. (**c**) The GSTC model

characterized by the superposition of waves radiated from different particles, each having a phase delay induced by the corresponding particle. As a result, the only hardware features of this model are the positions of the particles, i.e., STARS elements and their corresponding phase shifts associated with the transmission and reflection in Fig. 3.1a. The phase-shift model is widely adopted and convenient to use. However, it is an oversimplified representation of the actual physical process. As a result, it cannot accurately characterize either the energy flow at the surface or the nonlocal power transfer effects (Di Renzo et al., 2020).

3.2.2 The Load Impedance Model

In the load impedance model, each element is modeled as a lumped circuit having surface averaged electric and magnetic impedances of Y_m and Z_m. As illustrated in Fig. 3.1b, the physical processes of wave transmission and reflection may be portrayed as follows: Each element of the STARS is excited by the incident wave. After being excited, both electrical and magnetic currents are induced, whose intensity depends on the effective voltage of the incident wave and the equivalent load impedance of the circuit element. Finally, the currents induced generate an EM field, which is radiated toward both sides of the STARS. As a result, the hardware features of the model are the position and load impedances of each passive element. The problem of determining the field radiated by the currents flowing through each element is left to be dealt with by the channel model. The load impedance model can be reduced to the phase-shift model by incorporating some further idealized simplifying assumptions because both the TRCs of the surface can be formulated as a function of the surface impedances. Specifically, the TRCs of the m-th element are defined as the ratio between electric fields, which may be represented by complex numbers (Xu et al., 2021). The argument of the TRCs for each element corresponds to the phase delay values in the phase-shift hardware model. In light of this, the phase-shift model can be regarded as a simplified version of the load impedance model.

3.2.3 The Generalized Sheet Transition Conditions Model

The generalized sheet transition conditions (GSTC) model (Di Renzo et al., 2020) is the most general one of the three hardware models discussed because it is based on a continuous distribution of the electric and magnetic polarization densities of the surface, instead of relying on a finite number of impedance values. The GSTC model uses the electric and magnetic susceptibilities as a function of position on the surface for characterizing the smart surface. According to Maxwell's equations, the conventional boundary conditions at the surface describe the discontinuity of the EM field in terms of the surface electric and magnetic polarization densities. In the GSTC model, these surface electric and magnetic polarization densities are induced by the incident field and depend on the polarizability densities of the material. As illustrated in Fig. 3.1c, the transmitted and reflected fields can be formulated using only material-dependent surface susceptibility dyadics, $\chi_{ab}(x, y)$, as a function of the position on the STARS (Di Renzo et al., 2020). Moreover, the surface susceptibility dyadics are determined by the electric and magnetic polarizabilities of the scatterers, which are microscopic properties of the material. Thus, the GSTC model is capable of describing the metasurface-based STARS implementations relying on small periodic structures. At the same time, the GSTC model can also be used for modeling patch-array-based implementations by taking the surface average of the electric and magnetic polarizabilities within each element.

3.2.4 Summary and Outlook

In Table. 3.2, we summarize the characteristics of the three STARS hardware models discussed. The phase-shift and load impedance models best represent the patch-array-based implementations, while the GSTC model accurately mimics the metasurface-based implementations. At the time of writing, the existing papers on modeling and analyzing STARSs have adopted only the phase-shift model, which is a far-field, ray-optics approximation of the actual physical process. Future directions for STARS channel hardware modeling include proposing more physics-compliant models exploiting the above-mentioned load impedance and GSTC models. Moreover, based on different STARS hardware implementations, the correlation between the phase shifts of the transmitted and reflected signals should also be considered during hardware modeling.

Table 3.2 Comparing different hardware models for STARSs

Hardware models	Properties used for modeling	Apply to	Advantages	Disadvantages
Phase-shift model	Phase-shift (delay) values	Patch-array-based	Compact and easy to use	Oversimplified
Load impedance model	Surface averaged impedances	Patch-array-based	Compact and accurate	Not general
GSTC model	Surface susceptibility dyadics	Metasurface-based	General and accurate	Complicated

3.3 Channel Models Based on EM Arguments

Once the hardware model has been selected, the final step to reconcile physical implementations with communication theory is to adopt a channel model for determining the received signal power. There are various physics-based approaches that can be used to develop channel models for STARSs (Danufane et al., 2021; Booker & Clemmow, 1950; Gradoni & Di Renzo, 2021); hence below we critically appraise five promising approaches.

3.3.1 Far-Field Channel Models

One of the most widely accepted far-field channel models is the ray-tracing-based one (Xu et al., 2021). The ray-tracing technique has long been adopted as an efficient way to simplify the calculation of wave propagation and obtain the channel gains of Rxs located within the far-field region. In addition, the conventional ray-tracing technique is only compatible with the phase-shift-based STARS hardware model or the load impedance model. This is because the ray-tracing technique assumes a finite number of scatters and studies their sum, and hence it cannot deal with continuous phase profiles.

In the context of analyzing the STARS, the ray-tracing technique relies on the following assumptions:

1. Each element of the STARS is treated as a distinct scatter having a known location and dielectric properties.
2. The wave impinging on the STARS is regarded as a collection of rays, each falling on a single element. Thus, the received bundle of rays is constituted by a discrete 2D array instead of an EM field in three-dimensional (3D) space.
3. The interactions between each ray and each element, including the transmission and reflection, are studied using geometrical optics instead of wave optics.

3.3.2 *Near-Field Channel Models*

The Huygens-Fresnel principle is a powerful method of solving the problems of wave propagation in both the far-field and near-field regions of the STARS. It states that every point on a wavefront is itself the source of spherical wavelets, and the sum of these spherical wavelets forms the wavefront. The 2D wavefront should preferably be located in the same plane as the STARS, and the contribution to the received field at each wave point is proportional to its area on the wavefront, the amplitude of the corresponding wavelet, the leaning factor at each point on the wavefront, and to the reciprocal of the distance between each point on the wavefront and the Rx.

The Huygens-Fresnel principle-based near-field channel model is the best-suited one for studying the wavefront transformation function of the STARS. For example, if the incident field is a plane wave, then its wavefront evaluated at any plane should be of a uniform amplitude associated with a linear phase profile. Suppose we adopt the simplest phase-shift STARS hardware model, and then a wavefront transformation from a plane wave to other plane waves can be realized by implementing a linear phase gradient for both TRCs. Despite all the numerous advantages of the Huygens-Fresnel principle-based channel models, they are only best suited for describing the free-space systems. For wireless networks involving both STARSs and other uncontrollable scatters in the environment, we have to rely on other channel models.

3.3.3 *Other Channel Models*

3.3.3.1 Green's Function Method-Based Channel Models

The Green's function method is a mathematical procedure of solving inhomogeneous linear differential equations. According to Maxwell's equations, the EM field on both sides of the STARS satisfies the inhomogeneous Helmholtz equation. The Helmholtz equation represents the wave function that links the spatial derivative and the time derivative of the field with the source. Additionally, for fully characterizing the field, a boundary condition (Danufane et al., 2021) is needed. Depending on the type of Rx analyzed, the closed surface boundary should enclose the space, where the target Rx is located. This closed surface is formed by an infinitely large plane at the smart surface and a hemisphere having a radius tending to infinity. The boundary condition can be expressed as the complex value of the EM field at the surface, which is known as the Dirichlet boundary condition, or as the derivative of the EM field along the direction perpendicular to the surface, which is known as the Neumann boundary condition. The critical challenge of Green's function-based channel model is to choose the appropriate boundary conditions since all the information carried by the STARS can only be conveyed through the boundary conditions. In light of this, Green's function-based channel models are more compatible with the GSTC hardware model since they both are capable of characterizing detailed properties of the smart surface.

3.3.3.2 Angular Spectrum-Based Channel Models

They also rely on the Huygens-Fresnel principle. However, instead of calculating the channel gain using the Fresnel-Kirchhoff diffraction formula, the model exploits the fact that the EM fields on both sides of the STARS having any arbitrary distributions can be regarded as a collection of plane waves travelling in different directions. In light of this, provided that the spectrum of this collection of plane waves can be determined, the channel gain can also be derived at any point. In angular spectrum-based channel models, the wavefront is chosen as the plane in which the STARS lies. In the field of antenna design, the 2D spatial distribution of the EM field at this chosen wavefront is referred to as the *aperture distribution* (Mailloux, 2017), characterizing the complex-valued amplitude as a function of position on the smart surface. An essential statement that empowered the angular spectrum-based channel models is that the plane-wave angular spectrum of an arbitrary waveform can be given by the Fourier transform of the aperture distribution (Booker & Clemmow, 1950). In general, an arbitrary aperture distribution can be expressed as a Fourier expansion of a series of plane waves having different wave numbers. Correspondingly, the radiation obeying this aperture distribution can be expressed as a spectrum of plane waves. More detailed inspection shows that not all the wave numbers in the Fourier expansion of the aperture distribution correspond to propagating waves. In fact, aperture distributions having large wave number values give rise to *evanescent waves*. These waves do not propagate well, because they decay exponentially with distance, and hence they do not usefully contribute to the received power beyond a few wavelengths. The field region affected by these evanescent waves is termed as the reactive near-field region.

3.3.3.3 Equivalent Circuit-Based Channel Models

Instead of studying the propagation or generation of the EM wave, the equivalent circuit-based channel models characterize the channel between the STARS and the Rxs using a linear transformation (Gradoni & Di Renzo, 2021). Explicitly, in this model, each element and Rx is represented by specific *ports* of the circuit. The overall circuit consists of a collection of load impedances. The total number of ports is equal to the number of STARS elements plus the number of receiving antennas. Each port has different voltages and hence carries different currents. The load impedance matrix relates these voltages and currents to each other by a linear transformation. Thus, at the receiving ports, the time-averaged power can be calculated as the product of the current and voltage. However, in general cases, the

EM field response of the elements is not linear. As a result, we believe that the validity of the model in general scenarios has to be further justified.

The advantages and disadvantages of the above five channel models for STARSs are summarized in Table 3.3.

3.4 Dual-Sided STARS

3.4.1 EM-Based Modeling

In Chap. 2, two complex coefficients, namely TRCs, are introduced to characterize the simultaneous transmission and reflection features of each element. However, these two coefficients are derived by assuming that the signals only incident from one side of the surface and there is no signal incident from the other side. The signal model is generally unknown for the case where there are wireless signals simultaneously incident on both sides of the STARS, namely *dual-sided incidence.* To fill this knowledge gap, in this section, we provide a signal model for *dual-sided* STARSs. Since the "STAR" process occurs on each side, we introduce two pairs of coefficients in the following general signal model.

Theorem 3.1 *As illustrated in Fig. 3.2, for vertically polarized wireless signals, the signals radiating from the m-th element toward Space A (s'_A) and toward Space B (s'_B) are linear superposition of the incident signals ($s_{A/B}$):*

$$s'_A = \sum_{\theta \in \Theta_A} R_m^A(\theta) \cdot s_A(\theta) + \sum_{\theta \in \Theta_B} T_m^{B,A}(\theta) \cdot s_B(\theta), \tag{3.1}$$

$$s'_B = \sum_{\theta \in \Theta_A} T_m^{A,B}(\theta) \cdot s_A(\theta) + \sum_{\theta \in \Theta_B} R_m^B(\theta) \cdot s_B(\theta), \tag{3.2}$$

where $T_m^{A,B}(\theta)$ and $T_m^{B,A}(\theta)$ are the transmission coefficients (TCs) from Space A/B to Space B/A, $R_m^{A/B}(\theta)$ are the reflection coefficients (RCs) for Space A/B, subscript m denotes the m-th STARS element, θ is the incidence angle of the signal, and Θ_A and Θ_B are the set of angles of the incident signals from Space A and Space B, respectively.

Proof The key for this proof is to demonstrate that the received signals are the linear superposition of the incident signals from different directions. According to Estakhri and Alu (2016), for STARS that is electrically thin, the EM response can be equivalently described using surface electric and magnetic currents $\mathbf{J}_s$ and $\mathbf{K}_s$. Note that the direction of $\mathbf{K}_s$ is determined through the curl right-hand rule, which is illustrated in Fig. 3.3. Considering passive STARS elements, these currents are excited by the total EM field $\mathbf{E}_{tot}$, $\mathbf{H}_{tot}$, and scaled by the electric admittance ($\overline{\overline{\mathbf{Z}}}_{\mathrm{e}}$) and magnetic impedance ($\overline{\overline{\mathbf{Z}}}_{\mathrm{m}}$) tensors at each element, i.e., $\mathbf{J}_s = \overline{\overline{\mathbf{Z}}}_{\mathrm{e}} \cdot \mathbf{E}_{tot}$ and

Table 3.3 Comparing different channel models for STARSs

Channel models	Advantages	Disadvantages	Challenges
Ray tracing	Simplistic in form and easy to calculate channel gain	Apply only under certain conditions in far-field regions	Describing the channel in the near-field regions
Huygens-Fresnel principle	Fundamental and applies to near-field regions	Apply only to free-space scenarios or LoS-dominate links	Choosing proper wavefront to characterize the STARS
Angular spectrum	Convenient for designing desired aperture distributions	Apply only to free-space scenarios or LoS-dominate links	Deciding the boundaries between different field regions
Equivalent circuit	Simplistic in form and easy to calculate channel gain	Apply only for linear system in free space	Describing the channel for system with nonlinear filters
Green's function method	Fundamental and apply to general cases	Complex and requires detailed system specifications	Proper boundary conditions to characterize the environment

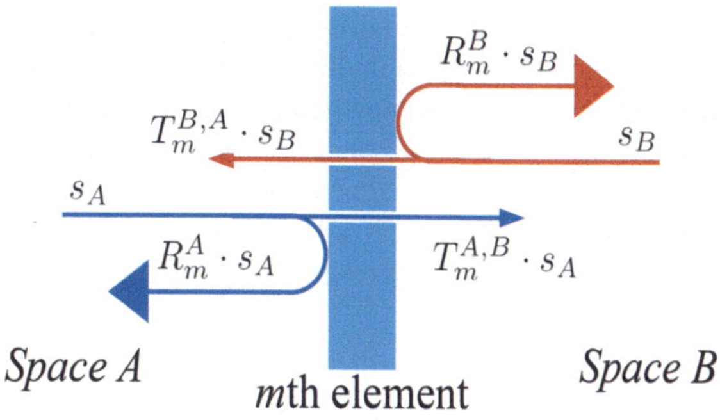

Fig. 3.2 A general signal model for dual-sided STARSs

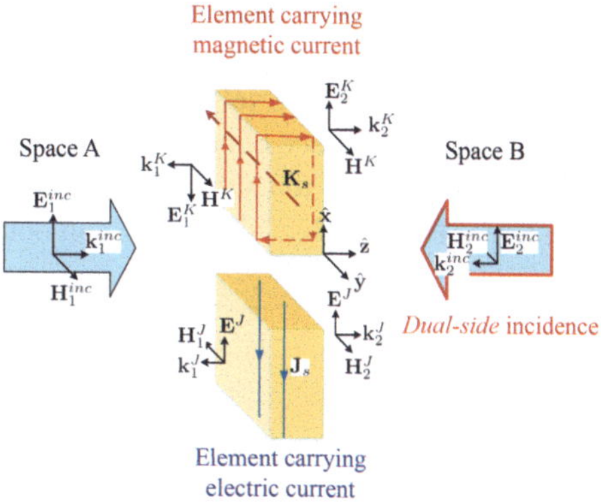

Fig. 3.3 Schematic of the STARS element under EM wave incidence

$\mathbf{K}_s = \overline{\overline{\mathbf{Z}}}_\mathrm{m} \cdot \mathbf{H}_{tot}$. As illustrated in Fig. 3.3, the electric current density $\mathbf{J}_s$ generates fields $\mathbf{E}^J$, $\mathbf{H}_1^J$, and $\mathbf{H}_2^J$, while the magnetic current density $\mathbf{K}_s$ generates fields $\mathbf{E}_1^K$, $\mathbf{E}_2^K$, and $\mathbf{H}^K$. The magnitude of the radiating field, $\mathbf{E}^J$ and $\mathbf{E}^K$, are proportional to the strength of $\mathbf{J}_s$ and $\mathbf{K}_s$. Thus, for signals with the same polarization, the received field can be expressed as the linear combination of the incident fields $\mathbf{E}_1^{inc}$ and $\mathbf{E}_2^{inc}$. Finally, by using the fact that the signal $s_{A/B}$ is proportional to the magnitude of the corresponding electric field, i.e., $s_{A/B} = ||\mathbf{E}_{A/B}^{inc}||_2$, the theorem is derived. □

3.4.2 *Signal Modeling for Symmetrical Dual-Sided STARS Elements*

For a general scenario, the TRCs for the two sides can be different, i.e., $\tilde{T}_m^{A,B} \neq T_m^{B,A}$ and $\tilde{R}_m^A \neq \tilde{R}_m^B$. However, for the case where the surface impedances can be characterized with scalars (Estakhri & Alu, 2016), the elements have symmetrical EM response for the two directions, i.e., $\tilde{T}_m^{A,B} = \tilde{T}_m^{B,A}$ and $\tilde{R}_m^A = \tilde{R}_m^B$. We separately study the radiation of electric and magnetic currents.[1] The schematic illustration of the STARS under EM incidence on both sides is presented in Fig. 3.3. In *Space A/B*, the total electrical radiation comes from the contributions of these induced electric fields and the overall incident fields $\mathbf{E}_1^{inc}$ and $\mathbf{E}_2^{inc}$. The strengths of the surface electric and magnetic currents are proportional to the sum of the incident and radiated fields at the STARS (Zhu et al., 2014). Thus, the density of the electric and magnetic currents are respectively given by

$$\mathbf{J}_s = (\mathbf{E}_1^{inc} + \mathbf{E}_2^{inc} + \mathbf{E}^J)/Y_m, \tag{3.3}$$

$$\mathbf{K}_s = (\mathbf{H}_1^{inc} - \mathbf{H}_2^{inc} + \mathbf{H}^K)Z_m, \tag{3.4}$$

where Y_m and Z_m are the scalar electric and magnetic impedance of the m-th element, respectively. To solve the generated electric fields $\mathbf{E}^J$ and $\mathbf{E}^K$, we consider the following boundary conditions. The boundary conditions (Rothwell & Cloud, 2018) of EM field at the surface can be expressed as $\hat{z} \times (\mathbf{H}_2^J - \mathbf{H}_1^J)\hat{y} = \mathbf{J}_s\hat{x}$ and $\hat{z} \times (\mathbf{E}_1^K - \mathbf{E}_2^K)\hat{x} = \mathbf{K}_s\hat{y}$. Thus, we can obtain the following:

$$\mathbf{E}^J = -\frac{\eta(\mathbf{E}_1^{inc} + \mathbf{E}_2^{inc})}{2Y_m + \eta}, \tag{3.5}$$

$$\mathbf{E}_1^K = -\mathbf{E}_2^K = \frac{Z_m(\mathbf{E}_1^{inc} - \mathbf{E}_2^{inc})}{Z_m + 2\eta}, \tag{3.6}$$

where η is the free-space wave impedance.

For an Rx located in *Space A/B*, the received signal strength is proportional to the square of the overall received electric field. As illustrated in Fig. 3.3, for *Space A*, we collect all the wave vectors going toward the left, i.e., $\mathbf{k}_2^{inc}$, $\mathbf{k}_1^J$, and $\mathbf{k}_1^K$. Thus, the overall electric field propagating toward *Space A* through the m-th STARS element can be calculated as follows:

$$\mathbf{E}_A^{rec} = \mathbf{E}^J + \mathbf{E}_1^K + \mathbf{E}_2^{inc} = \underbrace{\left(\frac{Z_m}{Z_m + 2\eta} - \frac{\eta}{2Y_m + \eta}\right)}_{\tilde{R}_m^A}\mathbf{E}_1^{inc} + \underbrace{\left(\frac{2Y_m}{2Y_m + \eta} - \frac{Z_m}{Z_m + 2\eta}\right)}_{\tilde{T}_m^{A,B}}\mathbf{E}_2^{inc}. \tag{3.7}$$

Similarly, the overall electric field propagating toward *Space B* is as follows:

[1] Here, the magnetic current refers to the equivalent magnetic current generated by the vortex currents (Vágó, 1994).

$$\mathbf{E}_B^{rec} = \mathbf{E}^J + \mathbf{E}_2^K + \mathbf{E}_1^{inc} = \underbrace{\left(\frac{2Y_m}{2Y_m+\eta} - \frac{Z_m}{Z_m+2\eta}\right)}_{\tilde{T}_m^{B,A}} \mathbf{E}_1^{inc}$$

$$+ \underbrace{\left(\frac{Z_m}{Z_m+2\eta} - \frac{\eta}{2Y_m+\eta}\right)}_{\tilde{R}_m^{B}} \mathbf{E}_2^{inc}. \quad (3.8)$$

Next, exploiting (3.7) and (3.8), the signal model for dual-sided STARSs can be established and we have the following theorem.

Theorem 3.2 *For symmetrical dual-sided STARSs, the signals radiating from the m-th element toward Space A (s'_A) and toward Space B (s'_B) are as follows:*

$$s'_A = R_m \cdot s_A + T_m \cdot s_B, \quad (3.9)$$

$$s'_B = T_m \cdot s_A + R_m \cdot s_B, \quad (3.10)$$

where

$$T_m = \frac{2Y_m}{2Y_m+\eta} - \frac{Z_m}{Z_m+2\eta}, \quad (3.11)$$

$$R_m = \frac{Z_m}{Z_m+2\eta} - \frac{\eta}{2Y_m+\eta} \quad (3.12)$$

are the TRCs for each STARS element.

Proof The proof of this theorem is based on the boundary conditions and the density of the electric and magnetic currents given in (3.3) and (3.4). Exploiting these characteristics, the fields in both *Space A* and *Space B* were given in (3.7) and (3.8). Then, by using the fact that the received signal $s'_{A/B}$ is the proportional to the magnitude of the received electric field, i.e., $s'_{A/B} = ||\mathbf{E}_{A/B}^{rec}||_2$, (3.7) and (3.8) can be rewritten into the forms as in (3.9) and (3.10), respectively. □

3.5 Active STARSs

As discussed in the previous section, under dual-sided incidence, each STARS element has four TRCs, i.e., $\tilde{R}_m^A$, $\tilde{R}_m^B$, $\tilde{T}_m^{AB}$, and $\tilde{R}_m^{BA}$. These four coefficients transform the incident signals to the output signals of each STARS element. The relation between the incident and output signals is given by

$$\begin{pmatrix} y_m^A \\ y_m^B \end{pmatrix} = \begin{pmatrix} \tilde{R}_m^A & \tilde{T}_m^{AB} \\ \tilde{T}_m^{BA} & \tilde{R}_m^B \end{pmatrix} \begin{pmatrix} s_m^A \\ s_m^B \end{pmatrix}, \quad (3.13)$$

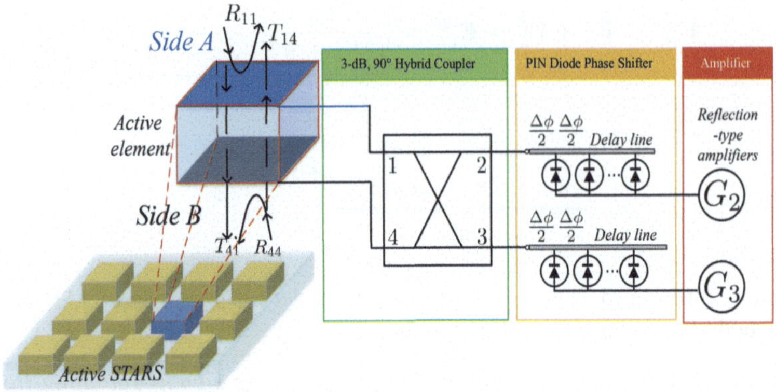

Fig. 3.4 Hardware structure of an active STARS element

or equivalently, $\mathbf{y}_m = \Xi_m \cdot \mathbf{s}_m$, where Ξ_m is the transmission and reflection matrix of the m-th STARS element. According to microwave network analysis, a STARS element is passive lossless if Ξ_m is unitary, i.e., $\Xi_m^H \Xi_m = \mathbf{I}_2$, where $\mathbf{I}_2$ denotes the two-by-two identity matrix. For passive lossy STARS element, $\Xi_m^H \Xi_m \prec \mathbf{I}_2$, and for active STARS element, $\Xi_m^H \Xi_m \succ \mathbf{I}_2$. In terms of reciprocity, a STARS can be either reciprocal or nonreciprocal. A STARS element is reciprocal if the transmission and reflection matrix Ξ_m is symmetrical, i.e., $\tilde{T}_m^{AB} = \tilde{T}_m^{BA}$. For nonreciprocal STAR matrix, we have $\tilde{T}_m^{AB} \neq \tilde{T}_m^{BA}$.

Next, we introduce the active STARS. As illustrated in Fig. 3.4, the proposed active STARS element consists of three major components: a quadrature hybrid, PIN diode phase shifters, and reflection-type amplifiers (Tohyama & Mizuno, 1979). Upon receiving the wireless signal, a 3-dB, 90° hybrid coupler with all four ports matched is employed to couple signals incident from both sides and minimize power loss. Then, PIN diode switches and phase shifters are connected to port 2 and port 3 of the hybrid coupler to impose adjustable phase delay to the signal. Finally, the delayed signals are fed to reflection-type amplifiers.

Depending on the configurations of the phase shifters and the gains of the amplifiers, the active STARS element structure has different overall transmission and reflection behavior. In the following, we formulate the TRCs of the active STARS element in terms of the phase-shift values of the PIN diode phase shifters and the gains of the reflective amplifiers.

3.5.1 *Hardware Model for Active STARSs*

3.5.1.1 Active STARS Element with Coupled Phase Shifts

If only one delay line and one reflection-type amplifier are available for each STARS element, the resultant element will exhibit coupled TRCs. As illustrated in Fig. 3.4, the delay line and amplifier can be connected to either port 2 or port 3 of the hybrid coupler. Without loss of generality, suppose that port 2 is connected to the reflection-type amplifier with gain G_2 and port 3 is grounded, i.e., $V_2^+ = \tilde{G}_2 V_2^-$ and $V_3^+ = 0$. Exploiting the scattering matrix of the quadrature hybrid, the output and input signal voltages have the following relation:

$$\begin{pmatrix} V_1^- \\ V_2^- \\ V_3^- \\ V_4^- \end{pmatrix} = -\frac{1}{\sqrt{2}} \begin{pmatrix} 0 & j & 1 & 0 \\ j & 0 & 0 & 1 \\ 1 & 0 & 0 & j \\ 0 & 1 & j & 0 \end{pmatrix} \begin{pmatrix} V_1^+ \\ V_2^+ \\ V_3^+ \\ V_4^+ \end{pmatrix}, \tag{3.14}$$

where $V_2^+ = \tilde{G} V_2^-$, $V_3^+ = 0$, and $\tilde{G} = G_2 \cdot e^{j\phi_2}$ is the complex amplitude response of the delay line and amplifier combined. Using the relation given in (3.14), the output signals through port 1 and port 4, i.e., the wireless signals exit from *side A* and *side B* of the STARS element, can be derived as follows:

$$V_1^- = -\frac{\tilde{G}}{2} \cdot V_1^+ + \frac{j\tilde{G}}{2} \cdot V_4^+ \tag{3.15}$$

$$V_4^- = \frac{j\tilde{G}}{2} \cdot V_1^+ + \frac{\tilde{G}}{2} \cdot V_4^+. \tag{3.16}$$

Theorem 3.3 *For STARS elements with coupled transmission and reflection phase shift, the four TRCs of the STARS elements that are connected to an amplifier with gain $\tilde{G}$ can be expressed as follows:*

$$R_m^A = \frac{V_1^-}{V_1^+}\bigg|_{V_4^+=0} = -\tilde{G}/2, \quad R_m^B = \frac{V_4^-}{V_4^+}\bigg|_{V_1^+=0} = \tilde{G}/2 \tag{3.17}$$

$$T_m^{AB} = \frac{V_1^-}{V_4^+}\bigg|_{V_1^+=0} = j\tilde{G}/2, \quad T_m^{BA} = \frac{V_4^-}{V_1^+}\bigg|_{V_4^+=0} = j\tilde{G}/2. \tag{3.18}$$

Proof The proof is straightforward by exploiting the definitions of the TRCs and the results obtained in (3.15) and (3.16). □

Remark 3.1 It can be observed that for the case where one amplifier is employed per element, the TRCs are highly coupled. Firstly, since the overall hardware structure is reciprocal, we have $T_m^{AB} = T_m^{BA}$. Next, consider the TRCs that are involved in the single-side incidence from *side A*, i.e., R_m^A and T_m^{BA}, and we have $(\theta_m^R - \theta_m^T)_A = \angle R_m^A - \angle T_m^{BA} = \pi/2$. Similarly, for *side B*, we have $(\theta_m^R - \theta_m^T)_B = \angle R_m^B - \angle T_m^{AB} = -\pi/2$. Note that the phase difference between the TRCs for the two sides will be exchanged if the amplifier is connected to port 3. This indicates that the active coupled phase-shift STARS follows the same transmission and reflection phase-shift correlation as the passive lossless STARSs, i.e., $\theta_m^R - \theta_m^T = \pm\pi/2$ (Xu et al., 2022). Finally, the amplitudes of all four TRCs are identical, and they are determined by the gain of the amplifier G_2.

3.5.1.2 Active STARS Element with Independent Phase Shifts

If two amplifiers are available for each STARS element, the resultant active STARS element has the ability to independently adjust the amplitude and phase shift of the TRCs. As illustrated in Fig. 3.4, the gains of the two amplifiers are G_2 and G_3. We further denote $\tilde{G}_2 = G_2 e^{j\phi_2}$ and $\tilde{G}_3 = G_3 e^{j\phi_3}$ as the combined gain and phase-shift delay of port 2 and port 3, respectively. By substituting $V_2^+ = \tilde{G}_2 V_2^-$ and $V_3^+ = \tilde{G}_3 V_3^-$ into (3.14), the output signals at port 1 and port 4 are given by

$$V_1^- = \frac{\tilde{G}_3 - \tilde{G}_2}{2} \cdot V_1^+ + \frac{j(\tilde{G}_2 + \tilde{G}_3)}{2} \cdot V_4^+ \tag{3.19}$$

$$V_4^- = \frac{j(\tilde{G}_2 + \tilde{G}_3)}{2} \cdot V_1^+ + \frac{\tilde{G}_2 - \tilde{G}_3}{2} \cdot V_4^+. \tag{3.20}$$

Theorem 3.4 *For STARS elements with independent active transmission and reflection phase shift, the four TRCs of the STARS elements that are connected to amplifiers with gain $\tilde{G}_2$ and $\tilde{G}_3$ can be expressed as follows:*

$$R_m^A = -R_m^B = (\tilde{G}_3 - \tilde{G}_2)/2 \tag{3.21}$$

$$T_m^{AB} = T_m^{BA} = j(\tilde{G}_2 + \tilde{G}_3)/2. \tag{3.22}$$

Proof *The proof is straightforward by exploiting the results obtained in* (3.19) *and* (3.20). □

Remark 3.2 It can be observed that since the amplitudes and phases of both G_2 and G_3 can be adjusted by tuning the phase-shift delay lines and amplifiers, the amplitudes and phases the TRCs can be independently configured. Moreover, the transmitting-only and reflecting-only modes can be achieved by letting $\tilde{G}_2 = \tilde{G}_3$ and $\tilde{G}_2 = -\tilde{G}_3$, respectively. However, due to the reciprocal design, the phase shifts and amplitudes of the TCs in the two directions remain identical, i.e., $T_m^{AB} = T_m^{BA}$.

3.5.2 System Model and Radiation Pattern

We investigate a two-user downlink communication system where a single-antenna base station (BS) is located on *side A* and the two users are located on different sides of the STARS. The direct link from BS to both users is assumed to be severely blocked. As a result, an active STARS is deployed to recover the service for the two users. Let g_m denote the channel between the BS and the m-th active STARS element, and h_m^χ denote the channel between the m-th active STARS element and the user χ, where $\chi \in \{A, B\}$ denotes the user on *side A* or *side B*. All channels are assumed to exhibit Rician fading, i.e., $|h_m^\chi| \sim \mathcal{R}(K_h^\chi, \Omega_h^\chi)$ and $|g_m| \sim \mathcal{R}(K_g, \Omega_g)$, where K_h^χ and K_g are the shape parameters, and Ω_h^χ and Ω_g are the scale parameters of the corresponding Rician channels. Considering the noise at the active STARS element and at each user, the received signal for user χ in the downlink scenario is given by

$$y^{\chi} = \sum_{m=1}^{M} h_m^{\chi} G_m^{\chi} e^{j\theta_m^{\chi}} (g_m s_m + v_m) + n^{\chi}, \tag{3.23}$$

where G_m^{χ} and θ_m^{χ} are the amplitude and phase shift of the TRCs of the m-th element corresponding to user χ, respectively. v_m is the noise caused by the m-th active STARS element, and n^{χ} is the noise caused by user χ. We further assume that the channel state information (CSI) is known at the active STARS. As a result, the cophase condition can be used to align the phase shift of each active STARS signal at both users.

In Fig. 3.5, we illustrate the angular patterns of the active STARS with coupled and independent transmission and reflection phase-shift and the passive lossless STARS as a baseline. In all scenarios, we consider a STARS with 18×18 elements. We assume that user A and user B are located at angular positions of 20° and 190° with equal distances of 10 m from the STARS, respectively. As can be seen from Fig. 3.5a, the coupled phase-shift active STARS forms a narrow beam toward user A. However, for the transmission space (the second and third quadrants), there are no observable beams. This is because due to the phase-shift coupling, only the reflection phase shift follows the cophase condition for user A. The radiation pattern for independent active STARS is shown in Fig. 3.5b. Compared with the coupled phase-shift scenario, the signal strength is focused for both users A and B. Moreover, based on the proposed hardware models, each independent active STARS element employs one more amplifier compared to the coupled phase-shift STARS element. As a result, the independent phase-shift STARS achieves larger amplitudes for both transmitted and reflected beams. In comparison, the passive lossless STARS shown in Fig. 3.5c has the smallest transmission and reflection beam amplitudes.

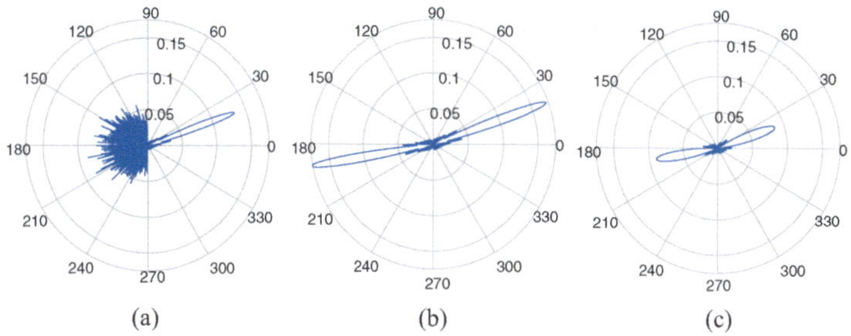

Fig. 3.5 Simulated radiation pattern of active and passive STARSs. (**a**) Coupled phase shift, active. (**b**) Independent phase shift, active. (**c**) Passive lossless STARS

3.6 Summary

In this chapter, the hardware implementation and modeling of STARSs were presented. Specifically, we pointed out that STARSs can be implemented by relying on both patch-array-based and metasurface-based technologies. Successful prototypes and hardware models of the STARS were also presented to illustrate how independent control of the transmitted and reflected signals can be achieved. Then, channel models with different levels of accuracy were summarized and compared. Finally, we provided the hardware model and signal model for the dual-sided STARSs and active STARSs.

References

Booker, H. G., & Clemmow, P. (1950). The concept of an angular spectrum of plane waves, and its relation to that of polar diagram and aperture distribution. *Proceedings of the IEE-Part III: Radio and Communication Engineering, 97*, 11–17.

Danufane, F. H., Renzo, M. D., de Rosny, J., & Tretyakov, S. (2021). On the path-loss of reconfigurable intelligent surfaces: An approach based on green's theorem applied to vector fields. *IEEE Transactions on Communications, 69*, 5573–5592.

Di Renzo, M., Zappone, A., Debbah, M., Alouini, M. S., Yuen, C., de Rosny, J., & Tretyakov, S. (2020). Smart radio environments empowered by reconfigurable intelligent surfaces: How it works, state of research, and the road ahead. *IEEE Journal on Selected Areas in Communications, 38*, 2450–2525.

Estakhri, N. M., & Alu, A. (2016). Wave-front transformation with gradient metasurfaces. *Physical Review X, 6*, 041008.

Gradoni, G., & Di Renzo, M. (2021). End-to-end mutual coupling aware communication model for reconfigurable intelligent surfaces: An electromagnetic-compliant approach based on mutual impedances. *IEEE Wireless Communications Letters, 10*, 938–942.

Mailloux, R. J. (2017). *Phased array antenna handbook*. Artech House.

Nemati, M., Maham, B., Pokhrel, S. R., & Choi, J. (2021). Modeling RIS empowered outdoor-to-indoor communication in mmwave cellular networks. *IEEE Transactions on Communications, 69*, 7837–7850.

NTT DOCOMO. (2020). DOCOMO conducts world's first successful trial of transparent dynamic metasurface. https://www.nttdocomo.co.jp/english/info/media_center/pr/2020/0117_00.html

Rothwell, E. J., & Cloud, M. J. (2018). *Electromagnetics*. CRC Press.

Tohyama, H., & Mizuno, H. (1979). 23-GHz band GaAs MESFET reflection-type amplifier. *IEEE Transactions on Microwave Theory and Techniques, 27*, 408–415.

Vágó, I. (1994). On the interface and boundary conditions of electromagnetic fields. *Periodica Polytechnica Electrical Engineering, 38*, 79–94.

Xu, J., Liu, Y., Mu, X., & Dobre, O. A. (2021). STAR-RISs: Simultaneous transmitting and reflecting reconfigurable intelligent surfaces. *IEEE Communications Letters, 25*, 3134–3138.

Xu, J., Liu, Y., Mu, X., Schober, R., & Poor, H. V. (2022). STAR-RISs: A correlated T&R phase-shift model and practical phase-shift configuration strategies. *IEEE Journal of Selected Topics in Signal Processing, 16*, 1097–1111.

Zhang, H., Zeng, S., Di, B., Tan, Y., Di Renzo, M., Debbah, M., Han, Z., Poor, H. V., & Song, L. (2022). Intelligent omni-surfaces for full-dimensional wireless communications: Principles, technology, and implementation. *IEEE Communications Magazine, 60*, 39–45.

Zhu, B. O., Chen, K., Jia, N., Sun, L., Zhao, J., Jiang, T., & Feng, Y. (2014). Dynamic control of electromagnetic wave propagation with the equivalent principle inspired tunable metasurface. *Scientific Reports, 4*, 1–7.

Zhu, W., Farmer, D. B., Jenkins, K. A., Ek, B., Oida, S., Li, X., Bucchignano, J., Dawes, S., Duch, E. A., & Avouris, P. (2013). Graphene radio frequency devices on flexible substrate. *Applied Physics Letters, 102*, 233102.

Part II
STARS-Aided Wireless Communications

Chapter 4
Performance Analysis for STARS

Abstract In this chapter, we evaluate the performance of STARSs under independent and coupled phase-shift models. We first introduce the most common performance metrics in the context of RIS-aided wireless communications. Then, for independent phase-shift STARSs, we analyze its diversity gain and compare it with conventional RISs. For coupled phase-shift STARSs, we propose three practical phase-shift configuration (PSC) strategies, namely, the primary-secondary PSC (PS-PSC), diversity-preserving PSC (DP-PSC), and T/R-group PSC (TR-PSC) strategies. Then, we derive the outage probability and power scaling law for each of the proposed PSC strategies. We also compare these strategies with the case of independent phase-shift STARS and STARS with random PSC. We further deduce the diversity orders of the users for all proposed strategies and show that, with the DP-PSC strategy, full diversity order can be simultaneously achieved for the users on both sides of the STARS.

4.1 Key Performance Metrics

In wireless communications, several performance metrics are used to evaluate the quality and efficiency of a communication system. Some of these metrics include bit error rate (BER), outage probability, diversity order, spectral efficiency, and capacity region. In the following, we first discuss these performance metrics in the context of RIS-aided wireless communications. Then, we provide the mathematical definitions of outage probability and diversity order that are chosen as the key performance metrics in our analysis.

4.1.1 Performance Metrics in the Context of RIS-Aided Wireless Communications

By examining different performance metrics in RIS-aided wireless communications, the impact of RIS deployment, channel characteristics, and system parameters

Y. Liu et al., *Simultaneously Transmitting and Reflecting Surfaces for Wireless Communications*, Wireless Networks, https://doi.org/10.1007/978-3-031-76848-4_4

on key performance indicators can be evaluated. This analysis aids in optimizing RIS configuration, power allocation, and transmission strategies to achieve higher reliability, improved spectral efficiency, and enhanced overall system capacity:

1. BER: BER quantifies the probability of bit errors in the received signal. In RIS-aided systems, BER reflects the accuracy of data transmission considering the effects of RIS-induced channel impairments, such as fading and reflections. Analyzing BER helps in evaluating the performance of modulation and coding schemes and assessing the system's error resilience.
2. Outage Probability: Outage probability refers to the probability that the received signal falls below a certain threshold, resulting in an outage or loss of communication. In RIS-aided systems, outage probability accounts for the impact of RIS deployment, channel fading, and interference. Analyzing outage probability helps in understanding the system's reliability and determining appropriate power allocation and resource allocation strategies.
3. Diversity Order: Diversity order characterizes the system's ability to combat fading and improve reliability by exploiting multiple independent signal paths. In RIS-aided systems, diversity order relates to the number of independent paths created by the RIS reflections, which can mitigate fading effects and enhance signal strength. A higher diversity order implies improved system performance in terms of error probability and outage resilience.
4. Spectral Efficiency: Spectral efficiency measures the amount of information that can be reliably transmitted over a given bandwidth. In RIS-aided systems, spectral efficiency considers the impact of RIS-induced beamforming and channel reflections on enhancing signal strength and reducing interference. Maximizing spectral efficiency enables higher data rates and improved utilization of available spectrum resources.
5. Capacity Region: The capacity region represents the set of all achievable rate tuples in a multiuser communication system. It characterizes the system's maximum achievable data rates for different users or channels. In RIS-aided systems, the capacity region takes into account the unique properties of RIS-aided communication, such as beamforming, interference mitigation, and channel enhancements. Analyzing the capacity region helps in optimizing resource allocation, user scheduling, and achieving efficient multiuser communication.

4.1.2 Outage Probability and Diversity Order

In the following analysis, we focus on outage probability and diversity order. The reasons are as follows: In RIS-aided systems, outage probability accounts for the impact of RIS deployment, channel fading, and interference. For STARSs, the phase-shift alignment of the TRCs is directly reflected in the user's outage probabilities. Furthermore, STARSs can create additional signal paths through simultaneous transmission and reflection, effectively increasing the diversity order. A higher diversity order implies improved system performance in terms of error probability and outage resilience. By analyzing the diversity order, we can understand the benefits of STARS-induced transmission/reflections in mitigating fading effects and enhancing signal strength:

- Outage Probability: The outage probability, denoted as P_{out}, is defined as the probability that the received signal power falls below a certain threshold γ_{th}. Mathematically, it can be expressed as

$$P_{\text{out}} = \Pr(\text{SNR} < \gamma_{\text{th}}). \tag{4.1}$$

- Diversity Order: The diversity order, denoted as d, quantifies the rate of decay of the error probability with respect to the transmit power or signal-to-noise ratio (SNR). The diversity order of a user is defined as follows Liu et al. (2017):

$$d = -\lim_{P_{BS}\to\infty} \frac{\log P_{\text{out}}}{\log P_{BS}}, \tag{4.2}$$

  where P_{BS} represents the transmit power. The diversity order d indicates the slope of the outage probability curve on a logarithmic scale, representing the system's ability to combat fading. A higher diversity order implies a faster decay of error probability with increasing SNR, indicating better system performance in terms of reliability and robustness against channel impairments.

4.2 Diversity Analysis for Independent Phase-Shift STARSs

To concisely highlight the benefits of STARSs, in this section, we present a comparison between the following two RISs with specific configurations:

- *STARS:* All STARS elements share common values for the power ratios of the transmitted signal and the reflected signal. However, the phase delays of each element can be independently configured.
- *Conventional RIS:* A composite smart surface consists of transmitting-only and reflecting-only RISs, each with M_t and M_r elements, respectively.

In the following, we compare the performance of the STARS with the conventional RIS and demonstrate the benefit of deploying STARSs in wireless systems. The power ratio between transmitted signals and the incident signal is fixed to β^T, and the power ratio of the reflected signal is fixed to β^R, for all STARS elements.[1] Based on the setup described in Sect. 2.1.3, we further assume that Tx and the

[1] It is worth mentioning that for STARSs, there is a trade-off between the received power of *Rx T* and *Rx R* by changing the power ratio of each element. Including these power ratios in the optimization can further improve the performance of the system.

STARS can be positioned with a line-of-sight (LoS) Tx-RIS link. Both Rxs are located in the far-field region of the RIS. The direct Tx-Rx links follow Ricean distribution as: $h^\chi \sim \mathcal{R}(K_d^\chi, \Omega_d^\chi)$, and the links between each STARS element and Rxs (r_m^χ) are identically and independently distributed, following Ricean distribution as: $r_m^\chi \sim \mathcal{R}(K_s^\chi, \Omega_s^\chi)$. Messages intended for different Rxs are sent on different orthogonal frequency bands. The signal sent by Tx can be expressed as $x = \sqrt{P_t}(\alpha^R s^R + \alpha^T s^T)$, where P_t is the transmit power, α^χ is the power allocation coefficient, and s^χ is the intended message for *Rx* χ. The SNR of Rx χ for the case of STARS or conventional RIS can be expressed as $\gamma_\mu^\chi = |H_\mu^\chi|^2 P_t(\alpha^\chi)^2/\sigma_0^2$, where $\mu \in \{S, C\}$ is an indicator representing "*S*" for the corresponding quantities in the STARS case, or "*C*" for the conventional RIS case, H_μ^χ is the overall channel, and σ_0^2 is the power spectrum density of the Gaussian white noise. According to the far-field channel model, the overall channel of Rx χ is

$$H_\mu^\chi = (\mathbf{r}^\chi)^H \mathbf{\Phi}_\mu^\chi \mathbf{h} + h^\chi, \tag{4.3}$$

where $\mathbf{\Phi}_S^T = \mathrm{diag}(T_1, \cdots, T_M)$, $\mathbf{\Phi}_S^R = \mathrm{diag}(R_1, \cdots, R_M)$, $\mathbf{\Phi}_C^T = \mathrm{diag}(0, \cdots, 0, T_1, \cdots, T_{M_t})$, and $\mathbf{\Phi}_C^R = \mathrm{diag}(R_1, \cdots, R_{M_r}, 0, \cdots, 0)$. For the case of STARS, under given transmit power P_t and target SNR γ_{target}^χ for Rx χ, the outage probability can be formulated as follows:

$$P_{out,S}^\chi = Pr\left\{\left|\sqrt{\beta^\chi}\sum_{m=1}^{M} r_m^\chi e^{j\phi_m^\chi} + h^\chi\right|^2 < \frac{\sigma_0^2 \gamma_{target}^\chi}{P_t(\alpha^\chi)^2}\right\}, \tag{4.4}$$

where β^T and β^R are the power ratios for the transmitted and reflected signals of all STARS elements. For the case of conventional RIS, the outage probability is

$$P_{out,C}^\chi = Pr\left\{\left|\sum_{m=1}^{M_\chi} r_m^\chi e^{j\phi_m^\chi} + h^\chi\right|^2 < \frac{\sigma_0^2 \gamma_{target}^\chi}{P_t(\alpha^\chi)^2}\right\}. \tag{4.5}$$

Note that the summation in (4.5) has M_t terms for *Rx T* and M_r terms for *Rx R*. In the following, we compare the maximum achievable diversity orders of Rxs for STARSs and conventional RISs.

Theorem 4.1 *For STARSs, the asymptotic outage probability of Rx χ can be expressed as follows (superscripts "χ" for K and Ω are omitted):*

$$P_{out,S}^\chi(\gamma_{target}^\chi) = \frac{2^{M+1}(K_s+1)^M(K_d+1)}{(2M+2)!\,\Omega_s^M\Omega_d(\alpha^\chi)^{2M+2}} \cdot (\beta^\chi)^{-M/2} e^{-MK_s-K_d}\sigma_0^{2M+2}(\gamma_{target}^\chi)^{M+1}P_t^{-(M+1)}. \tag{4.6}$$

For conventional RISs, the asymptotic outage probability of Rx χ can be expressed as follows:

$$P_{out,C}^{\chi}(\gamma_{target}^{\chi}) = \frac{2^{M'+1}(K_s+1)^{M'}(K_d+1)}{(2M'+2)!\Omega_s^{M'}\Omega_d(\alpha^{\chi})^{2M'+2}}$$
$$\cdot e^{-M'K_s-K_d}\sigma_0^{2M'+2}(\gamma_{target}^{\chi})^{M'+1}P_t^{-(M'+1)}, \tag{4.7}$$

where $M' = M_t$ for Rx T and $M' = M_r$ for Rx R.

Proof Following the assumption of Ricean fading channels, the probability density functions (PDFs) of Tx-Rx direct link and RIS-Rx links can be expressed in the following series form:

$$p_{|r_m^{\chi}|}(x) = \frac{2(K_s^{\chi}+1)}{\sqrt{\beta^{\chi}\Omega_s^{\chi}}}e^{-K_s^{\chi}}x + o(x^2), \tag{4.8}$$

$$p_{|h^{\chi}|}(x) = \frac{2(K_d^{\chi}+1)}{\Omega_d^{\chi}}e^{-K_d^{\chi}}x + o(x^2), \tag{4.9}$$

where $o(\cdot)$ is the little-O notation, and $o(x^2)$ denotes a function that is asymptotically smaller than x^2. To obtain the PDF of the magnitude of the overall channel ($|H_k^{\mu}|$), we calculate its Laplace transform according to the convolution theorem. Considering the case of STARS, we have

$$\mathcal{L}\{p_{|H_S^{\chi}|}(x)\}(t) = \left(\mathcal{L}\{p_{|r_m^{\chi}|(x)}\}\right)^M \mathcal{L}\{p_{|h^{\chi}|}(x)\}$$
$$= \frac{2^{M+1}(K_s+1)^M(K_d+1)}{(\beta^{\chi})^{M/2}\Omega_s^M\Omega_d}e^{-MK_s-K_d}\cdot t^{-2M-2}. \tag{4.10}$$

Thus, by applying the inverse Laplace transform, the outage probability can be calculated according to

$$P_{out,S}^{\chi} = \int_0^{\frac{\sigma_0}{\alpha^{\chi}}\sqrt{\frac{\gamma_{target}^{\chi}}{P_t}}} p_{|H_S^{\chi}|}(x)dx. \tag{4.11}$$

The proof for the case of conventional RISs follows the same strategy and thus is omitted here for brevity. □

In the following remark, we consider the diversity order defined through outage probability as follows:

$$d^{\chi} = -\lim_{P_t\to\infty}\frac{\log P_{out}^{\chi}(\gamma_{target}^{\chi})}{\log P_t}. \tag{4.12}$$

Remark 4.1 According to Theorem 4.1, the diversity orders of different Rxs can be readily obtained from (4.6) and (4.7). For STARSs, the maximum achievable diversity orders of both Rxs are $d_S^T = d_S^R = M + 1$. Thus, the combined diversity order is $d_S^T + d_S^R = 2M + 2$. For conventional RISs, the maximum achievable diversity order for *Rx T* and *Rx R* follows: $d_C^T + d_C^R = M + 2$, yielding smaller achievable diversity orders compared to STARSs.

4.3 Practical Phase-Shift Configuration Strategies for STARSs

4.3.1 System Model

To investigate the impact of the coupled phase-shift model given in (2.11) on wireless communication performance and to reveal fundamental insights for system design, we consider a STARS-aided two-user downlink network, as illustrated in Fig. 4.1. In particular, we assume a single-antenna BS serves two single-antenna users, which are located on different sides of the STARS. The user located in the transmission region of the STARS is referred to as user T, while the user located in the reflection region is referred to as user R. The channel between the BS and the m-th STARS element is denoted by g_m, and the channels between the m-th STARS element and user T and user R are denoted by h_m^T and h_m^R, respectively. Let $h_d^{T/R}$ denote the direct link between the BS and user T/R. All channels are assumed to

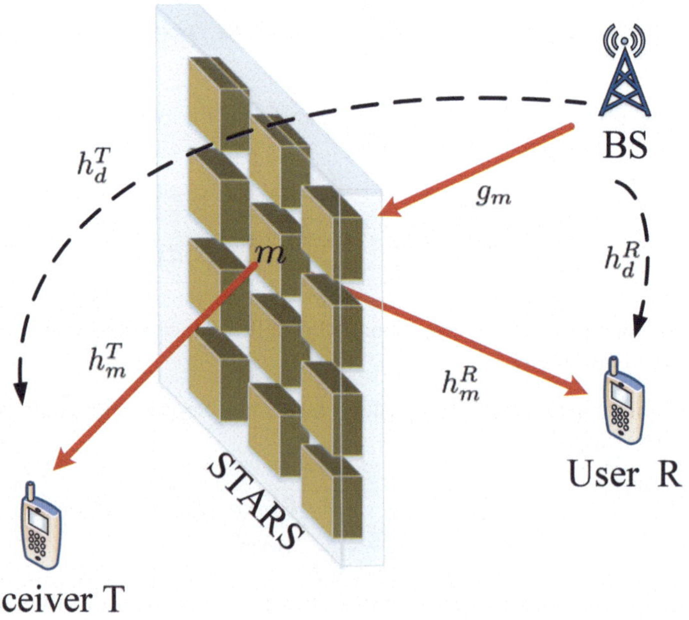

Fig. 4.1 System model

be independent and identically distributed (i.i.d.) Rician fading channels.[2] Thus, for $\chi \in \{T, R\}$, indicating user T/R, h_m^χ can be expressed as follows:

$$h_m^\chi = \sqrt{\frac{\rho_0^\chi}{(d_m^\chi)^\alpha}} \left(\sqrt{\frac{K_h^\chi}{K_h^\chi + 1}} h_m^{\mathrm{LoS},\chi} + \sqrt{\frac{1}{K_h^\chi + 1}} h_m^{\mathrm{NLoS},\chi} \right), \tag{4.13}$$

where d_m^χ denotes the distance between the m-th STARS element and user χ, α is the path loss exponent, K_h^χ denotes the Rician factor, ρ_0^χ is the path loss at a reference distance of 1 meter, $h_m^{\mathrm{LoS},\chi}$ is the LoS component, and $h_m^{\mathrm{NLoS},\chi}$ is the non-line-of-sight (NLoS) component that is Rayleigh fading. Channels g_m and h_d^χ are modeled similarly to (4.13) with Rician factors K_g and K_d^χ, respectively.

The performance of STARS-aided networks depends on the choice of the phase shifts (θ_m^T, θ_m^R) and the amplitudes (β_m^T, β_m^R) applied by each element. This is because the end-to-end channel gain between the BS and user χ is given by

$$H^\chi = \sum_{m=1}^{M} g_m h_m^\chi \beta_m^\chi e^{j\theta_m^\chi} + h_d^\chi, \ \chi \in \{T, R\}. \tag{4.14}$$

For ease of presentation, we rewrite the overall channel gain by separating the amplitude terms and the phase-shift terms as follows:

$$H^\chi = \sum_{m=1}^{M} \beta_m^\chi |g_m||h_m^\chi| \exp\{j(\angle g_m + \angle h_m^\chi + \theta_m^\chi)\} + h_d^\chi, \tag{4.15}$$

where $\angle \xi$ denotes the argument (complex angle) of a complex number ξ.

In the following, we propose three STARS phase-shift configuration (PSC) strategies for the coupled phase-shift model in (2.11), namely the primary-secondary (PS)-PSC, the diversity-preserving (DP)-PSC, and the T/R-Group (TR)-PSC strategies. We note that due to the T&R phase-shift correlation in (2.11), existing results on phase-shift optimization for conventional RISs and STARSs with independent phase shifts are not applicable for the considered passive STARS.

4.3.2 Primary-Secondary Phase-Shift Configuration (PS-PSC) Strategy

For the PS-PSC strategy, we adjust the phases of the TRCs, while assuming that the power ratio between the amplitudes of the TRCs is fixed to the same value

[2] According to Marzetta and Ngo (2016), i.i.d. fading can be achieved by employing uniform linear arrays (ULAs) with $\lambda/2$-separation between adjacent antennas.

for each STARS element. The STARS can be optimized for enhancing the channel gain of one user, referred to as the primary user, while simultaneously serving the other user, referred to as the secondary user. Based on this strategy, the PSC of the STARS elements can be determined in the following two steps. Without loss of generality, assume that user R is the primary user. In the first step, the phase terms $\{\theta_1^R, \cdots, \theta_M^R\}$ are configured to maximize the channel gain of user R without considering the phase-shift correlation constraint in (2.11). This yields the following optimization problem:

$$\max_{\theta_1^R,\cdots,\theta_M^R} |H^R|^2, \text{ s.t. } \theta_m^R \in [0, 2\pi), \ \forall m = 1, 2, \cdots, M. \tag{4.16}$$

In the second step, the $\{\theta_1^T, \cdots, \theta_M^T\}$ are optimized to maximize the overall channel gain of user T for given $\{\theta_1^{R^*}, \cdots, \theta_M^{R^*}\}$ taking into account the phase-shift correlation model in (2.11). In this case, the optimization of ϕ_m^T under constraint (2.11) is equivalent to optimizing the auxiliary bits $\{\nu_1, \cdots, \nu_M\}$ of the STARS, which leads to the following optimization problem:

$$\max_{\nu_1,\cdots,\nu_M} |H^T|^2, \text{ s.t. } \nu_m = 0 \text{ or } 1, \ \forall m = 1, 2, \cdots, M. \tag{4.17}$$

Since constraint (2.11) is not taken into account for the maximization with respect to $\theta_1^R, \cdots, \theta_M^R$ in (4.16), the optimal values of both $\theta_1^R, \cdots, \theta_M^R$ and $\nu_1, \cdots, \nu_M$ can be obtained in closed form if the CSI is known. According to the cophase condition (Liu et al., 2021), the optimal solution to problem (4.16) is given by

$$\theta_m^{R^*} = (\angle h_d^R - \angle h_m^R - \angle g_m) \mod 2\pi. \tag{4.18}$$

For the secondary user, the phase shift according to the cophase condition would follow as $\theta_m^{T'} = (\angle h_d^T - \angle h_m^T - \angle g_m) \mod 2\pi$. However, this phase-shift value might not be possible because of (2.11) for the given $\theta_m^{R^*}$. Thus, we obtain the optimal choice of the auxiliary bits as follows:

$$\nu_m^* = \begin{cases} 0 & \text{if } \theta_0^T \leq \theta_m^{R^*} < \theta_0^T + \pi, \\ 1 & \text{if } \theta_0^T - \pi \leq \theta_m^{R^*} < \theta_0^T, \end{cases} \tag{4.19}$$

where $\theta_0^T = \angle h_d^T$ is the phase of the direct link between user T and the BS. The PSC in (4.19) effectively selects the θ_m^T value that is closer to $\theta_m^{T'}$ given the two possible choices, i.e., $\theta_m^T = \theta_m^{R^*} - \pi/2$ and $\theta_m^T = \theta_m^{R^*} - 3\pi/2$.

Table 4.1 DP-PSC strategy, $\theta'_m = (\Delta\theta_m^R + \Delta\theta_m^T)/2$

$(\Delta\theta_m^R - \Delta\theta_m^T) \in$	$[0, \pi)$	$[\pi, 2\pi)$	$[-\pi, 0)$	$(-2\pi, -\pi)$
θ_m^{T*}	$\theta'_m - \frac{\pi}{4}$	$\theta'_m - \frac{3\pi}{4}$	$\theta'_m + \frac{\pi}{4}$	$\theta'_m + \frac{3\pi}{4}$
ν^*	0	1	1	0
θ_m^{R*}	$\theta'_m + \frac{\pi}{4}$	$\theta'_m + \frac{3\pi}{4}$	$\theta'_m - \frac{\pi}{4}$	$\theta'_m - \frac{3\pi}{4}$

4.3.3 Diversity-Preserving Phase-Shift Configuration (DP-PSC) Strategy

In the previous section, we showed that one of the benefits of STARSs is that they can achieve full diversity orders for both the user T and the user R. Here, we show that this is possible even for STARSs with coupled phase shifts. To this end, we propose a DP-PSC strategy. For given amplitude coefficients, the phase shifts and the values of the auxiliary bits are determined as follows: First, according to the cophase condition, we calculate the phase-shift value that would maximize the channel gains of both users without taking into account the phase-shift correlation constraint. This leads to

$$\Delta\theta_m^\chi = (\angle h_d^\chi - \angle h_m^\chi - \angle g_m) \mod 2\pi. \tag{4.20}$$

Then, taking the phase correlation into account, the PSC for the DP-PSC strategy is determined based on the calculated phase-shift values $\Delta\theta_m^R$ and $\Delta\theta_m^T$ and is given in Table 4.1 along with the corresponding auxiliary bit ν. To illustrate the idea behind this PSC, let $\delta_m^\chi = |\theta_m^{\chi^*} - \Delta\theta_m^\chi|$ denote the phase error for user χ. The PSC given in Table 4.1 imposes identical phase errors on both users, i.e., $\delta_m^R = \delta_m^T$. Moreover, it can be verified that $\delta_m^\chi \leq \pi/4$. This property is crucial for the DP-PSC strategy to exploit the spatial diversity of the STARS.

4.3.4 T/R-Group Phase-Shift Configuration (TR-PSC) Strategy

For the proposed TR-PSC strategy, the elements of the STARS are partitioned into two groups. The elements in the T group employ full transmission and serve only the user T, while the elements in the R group employ full reflection and serve only the user R. For the fully reflecting elements, we have $\beta_m^R = 1$ and $\beta_m^T = 0$, and for the fully transmitting elements, we have $\beta_m^T = 1$ and $\beta_m^R = 0$. Due to the decoupling of transmission and reflection at the element level, the phase correlation constraint in (2.11) does not apply to the TR-PSC strategy. Let $\mathcal{T}$ denote the set of indices of the elements in the T group. Then, the TRCs for the TR-PSC strategy are given by

$$\theta_m^R = \angle h_d^R - \angle h_m^R - \angle g_m, \; m \notin \mathcal{T}, \tag{4.21}$$

$$\theta_m^T = \angle h_d^T - \angle h_m^T - \angle g_m, \; m \in \mathcal{T}. \tag{4.22}$$

4.3.5 Summary and Discussion

In Fig. 4.2, we illustrate the three proposed PSC strategies for STARSs and the random PSC as a baseline scheme. At user χ, the overall phase of the signal received from element m is $\angle g_m + \angle h_m^\chi + \theta_m^\chi$. The small arrows in Fig. 4.2 indicate these overall phase terms. Under the assumption of i.i.d. fading channels, the yellow areas denote the possible ranges of the values of the overall phase terms. Note that these sectors are different in size for the different users and for different designs and indicate the possible deviation of the actual phase shift from the optimal values given by the cophase condition. As shown in Fig. 4.2a, the overall phase terms for the primary user under the PS-PSC strategy and for both users under the TR-PSC strategy are equal to the values obtained from the cophase condition. In contrast, the yellow sectors of the secondary user for the PS-PSC strategy have central angles of 180 degrees. The central angles of the yellow sectors of users under the DP-PSC strategy and the random PSC are 90 and 360 degrees, respectively. These deviations of the overall phase terms are different for different PSC strategies, which leads to significant performance difference for the users.

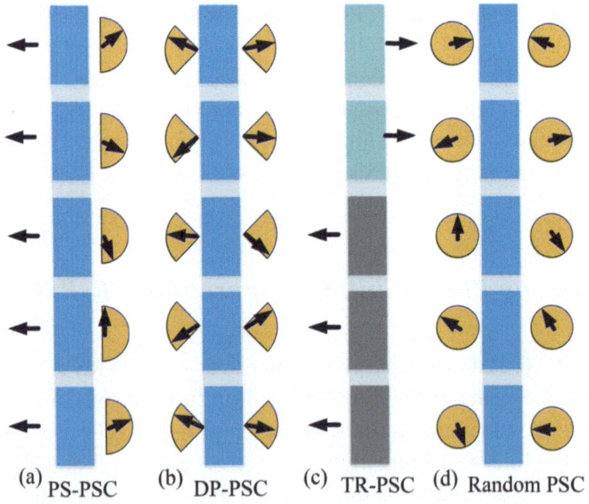

Fig. 4.2 Conceptual illustration of different STARS PSC strategies

4.4 Outage and Power Scaling Analysis

In the following, the communication performance of the T and R users is analyzed for the proposed PSC strategies, and performance lower and upper bounds are established. Specifically, we first derive the outage probabilities and diversity orders for the proposed PSC strategies. Then, we study the power scaling law, which quantifies how the average received power scales with the number of STARS elements.

4.4.1 *Multiuser Access Schemes*

Both orthogonal multiple access (OMA) and NOMA can be used for the BS to serve the users. In OMA, the BS serves both users in orthogonal frequency bands of equal size. In NOMA, the BS sends a superimposed signal to both users in the same time/frequency resource block. Let s^{χ} and c_{χ} denote the information symbol and the power allocation coefficient for user χ, respectively. In the following, we outline both considered multiple access schemes.

4.4.1.1 OMA

In this case, the BS communicates with both users in orthogonal frequency bands employing frequency-division multiple access (FDMA). Thus, the achievable rate for user χ is given by

$$R_{\chi}^{\mathrm{OMA}} = \frac{1}{2}\log_2\left(1 + \frac{P_{BS}c_{\chi}^2|H^{\chi}|^2}{\sigma_0^2/2}\right), \tag{4.23}$$

where P_{BS} and σ_0^2 are the transmit power of the BS and the variance of the additive white Gaussian noise (AWGN) at both users, respectively. In (4.23), we assume that the users employ orthogonal frequency bands of equal size. Thus, the bandwidth and the noise are both reduced by $1/2$.

For FDMA, the BS serves the users in two orthogonal frequency bands of equal sizes employing frequency-flat STARS TRCs. Suppose that for user χ the target data rate is $\tilde{R}_{\chi}$. Then, according to (4.23), the outage probability for the user is given by

$$P_{\mathrm{out},\chi}^{\mathrm{OMA}} = \Pr\{R_{\chi}^{\mathrm{OMA}} < \tilde{R}_{\chi}\} = \Pr\left\{|H^{\chi}|^2 < \frac{(2^{2\tilde{R}_{\chi}} - 1)\sigma_0^2}{2c_{\chi}^2 P_{BS}}\right\}, \tag{4.24}$$

where $\Pr\{A\}$ denotes the probability of event A. For ease of exposition, we define the following term as the OMA channel gain threshold for target data rate $\tilde{R}_\chi$:

$$\tau_\chi^{\mathrm{OMA}}(\tilde{R}_\chi) \triangleq (2^{2\tilde{R}_\chi} - 1)\sigma_0^2/(2c_\chi^2). \tag{4.25}$$

4.4.1.2 NOMA

For NOMA, one of the users employs successive interference cancelation (SIC) to remove interference. Specifically, assuming CSI-based user ordering, the user with the stronger channel first decodes the signal of the user with the weaker channel, before decoding its own signal. CSI-based user ordering can be carried out at the BS after CSI acquisition based on a suitable channel estimation method. For simplicity, we assume that user R has the stronger channel, i.e., $|H^R| \geq |H^T|$. To ensure that SIC is successfully carried out and to guarantee user fairness, the power allocation coefficients are set to ensure $c_R < c_T$. Therefore, user R will first detect the signal of user T via SIC, before detecting its own signal. The achievable rate for user R to detect the message of user T is given by

$$R_{R,T}^{\mathrm{NOMA}} = \log_2\left(1 + \frac{P_{BS}c_T^2|H^R|^2}{P_{BS}c_R^2|H^R|^2 + \sigma_0^2}\right). \tag{4.26}$$

Then, user R can detect its own signal after subtracting the signal of user T via SIC. Hence, the achievable rate of user R is given as follows:

$$R_{R,R}^{\mathrm{NOMA}} = \log_2\left(1 + \frac{P_{BS}c_R^2|H^R|^2}{\sigma_0^2}\right). \tag{4.27}$$

In contrast, user T directly detects its own signal by treating the signal of user R as interference. The corresponding achievable rate is given by

$$R_{T,T}^{\mathrm{NOMA}} = \log_2\left(1 + \frac{P_{BS}c_T^2|H^T|^2}{P_{BS}c_R^2|H^T|^2 + \sigma_0^2}\right). \tag{4.28}$$

Suppose that for NOMA user χ, the target data rate is $\tilde{R}_\chi$. Moreover, we consider the case where user R has the better channel condition. For given target data rates $\tilde{R}_T$ and $\tilde{R}_R$ of the users, their outage probabilities can be formulated as follows:

$$P_{\mathrm{out},R}^{\mathrm{NOMA}} = \Pr\left\{|H^R|^2 < \max\left\{\frac{(2^{\tilde{R}_T} - 1)\sigma_0^2}{[c_T^2 - (2^{\tilde{R}_T} - 1)c_R^2]P_{BS}}, \frac{(2^{\tilde{R}_R} - 1)\sigma_0^2}{c_R^2 P_{BS}}\right\}\right\}, \tag{4.29}$$

$$P_{\mathrm{out},T}^{\mathrm{NOMA}} = \Pr\Big\{|H^T|^2 < \frac{(2^{\tilde{R}_T}-1)\sigma_0^2}{[c_T^2-(2^{\tilde{R}_T}-1)c_R^2]P_{BS}}\Big\}. \tag{4.30}$$

For convenience, we define the following two NOMA channel gain thresholds:

$$\tau_R^{\mathrm{NOMA}} \triangleq \max\Big\{\frac{(2^{\tilde{R}_T}-1)\sigma_0^2}{c_T^2-(2^{\tilde{R}_T}-1)c_R^2}, \frac{(2^{\tilde{R}_R}-1)\sigma_0^2}{c_R^2}\Big\}, \tag{4.31}$$

$$\tau_T^{\mathrm{NOMA}} \triangleq \frac{(2^{\tilde{R}_T}-1)\sigma_0^2}{c_T^2-(2^{\tilde{R}_T}-1)c_R^2}. \tag{4.32}$$

Since both $\tilde{R}_R$ and $\tilde{R}_T$ appear in (4.29), for NOMA, user R will experience an outage if the target data rate of user T is chosen too large. This is because as user R applies SIC, it will declare an outage either when it cannot decode its own message or when it cannot decode the message of user T.

4.4.2 PS-PSC Strategy

For the primary user (user R in the considered case), the overall channel gain in (4.15) can be rewritten as follows:

$$H^R = \beta^R \sum_{m=1}^{M} |g_m||h_m^R| + h_d^R, \tag{4.33}$$

where β^R is the amplitude of the RC shared by all STARS elements, i.e., $\beta_m^R = \beta^R$, $\forall m = 1, 2, \cdots, M$, in (2.10).

4.4.2.1 Outage Probability

Given the overall channel gain in (4.33), we obtain the following result:

Theorem 4.2 *For the primary user (user R), the asymptotic outage probability for NOMA and OMA can be expressed as follows:*

$$P_{out}^R(\tau_R) = \frac{2^{M+1}(K_h^R+1)^M(K_g+1)^M(K_d+1)}{(2M+2)!\,\Omega_h^M\Omega_g^M\Omega_d^\chi}(\beta^R)^{-2M}$$
$$\cdot e^{-MK_h^R-MK_g-K_d^R}(\tau_R)^{M+1}P_{BS}^{-(M+1)}, \tag{4.34}$$

where $\tau_R = \tau_R^{NOMA}$ for NOMA and $\tau_R = \tau_R^{OMA}$ for OMA, and Ω_d^χ is the scale parameter of the direct link of user χ.

Proof See Appendix A. □

Next, we investigate the diversity order of user χ, which is defined as follows Liu et al. (2017):

$$d^\chi = -\lim_{P_{BS}\to\infty} \frac{\log P_{out}^\chi(\tau_\chi)}{\log P_{BS}}. \tag{4.35}$$

Corollary 4.1 *The diversity orders of the primary user for the PS-PSC strategy are identical for both NOMA and OMA. Exploiting* (4.34) *and using* (4.35)*, the diversity order for the primary user is given by*

$$d_{primary} = M + 1. \tag{4.36}$$

Proof Combining (4.34) with (4.35), it is straightforward to obtain (4.36). □

For the secondary user, the overall channel can be rewritten as follows:

$$H^T = \sum_{m=1}^{M} |g_m||h_m^T|e^{j\theta_m'}\beta^T e^{j\theta_m^T} + h_d^T = \beta^T \sum_{m=1}^{M} |g_m||h_m^T|e^{j(\theta_0^T+\delta_m)} + h_d^T, \tag{4.37}$$

where $e^{j\theta_m'}$ is the phase of $g_m h_m^T$, $\delta_m = \theta_m' + \theta_m^T - \theta_0^T$ is the residual phase, which satisfies $-\pi/2 < \delta_m < \pi/2$ according to (2.11) and (4.19), θ_0^T is the phase of the direct channel between user T and the BS, and β^T is the amplitude of the TC that is assumed to be equal for all STARS elements, i.e., $\beta_m^T = \beta^T = \sqrt{1-(\beta^R)^2}$, $\forall m = 1, 2, \cdots, M$. As a result, for the secondary user, the STARS elements can be regarded as a one-bit phase shifter. The asymptotic behavior and diversity order for one-bit phase shifters have been studied in Wang et al. (2020). Hence, given the presence of the direct BS-user link, the diversity order of the secondary user (user T) for both NOMA and OMA is given by[3]

$$d_{secondary} = \frac{M+3}{2}. \tag{4.38}$$

Remark 4.2 The diversity orders given in (4.36) and (4.38) include the contribution of the direct channel h_d^χ. For the scenario where the direct link is blocked, it can be shown that the diversity orders decrease by one, which leads to the following results:

[3] The diversity order of user R and user T switches if user T is the primary user.

$$d'_{\text{primary}} = M, \; d'_{\text{secondary}} = \frac{M+1}{2}. \tag{4.39}$$

4.4.2.2 Power Scaling Law

Due to the asymmetric PSC, the power scaling laws will be different for user R and user T. For convenience, let μ_h^X and σ_h^X denote the expected value and the variance of the amplitude of h_m^X, respectively. The following results hold Simon (2002)

$$\mu_h^X = \frac{1}{2}\sqrt{\frac{\pi \Omega_h^X}{(K_h^X + 1)}} L_{1/2}(-K_h^X), \tag{4.40}$$

$$\sigma_h^X = \Omega_h^X - (\mu_h^X)^2, \tag{4.41}$$

where $L_{1/2}(x)$ denotes the Laguerre polynomial. For the primary user, according to (4.33) and given that the $|h_m^R|$ are assumed to be i.i.d. Rician random variables, the expected value and variance of H^R can be expressed as follows:

$$\mathbb{E}[|H^R|] = |g|\beta^R \sum_m \mathbb{E}[|h_m^R|] + \mathbb{E}[|h_d^R|] = M\mu_h^R |g|\beta^R + \mathbb{E}[|h_d^R|], \tag{4.42}$$

$$\text{Var}[|H^R|] = M\sigma_h^R |g|^2 (\beta^R)^2 + \text{Var}[|h_d^R|], \tag{4.43}$$

where $|g| = \mathbb{E}[|g_m|]$ is the expected magnitude of the channel between the m-th STARS element and the BS.

For the secondary user, according to (4.37), the overall channel H^T is a complex-valued sum of the channel amplitude terms $|h_m^T|$ and the phase-shift terms $e^{j(\theta_0^T + \delta_m)}$. Due to the complexity of this mathematical expression, we first present a lemma as a building block for the power scaling analysis for the secondary user. We rewrite the overall channel of the secondary user as $H^T = h_s^T + h_d^T$, where $h_s^T = \beta^T \sum_{m=1}^M |g_m||h_m^T| e^{j(\theta_0^T + \delta_m)}$ is the STARS-aided channel.

Lemma 4.1 *The PDF of the magnitude of h_s^T can be approximated as follows:*

$$f_{|h_s^T|}(x) = \frac{x}{\beta^2} e^{-\frac{x^2+\alpha^2}{2\beta^2}} I_0\left(\frac{x\alpha}{\beta}\right), \tag{4.44}$$

where

$$\alpha = 2M\mu_h^T/\pi, \; \beta^2 = \frac{M}{2}\Omega_h^T, \tag{4.45}$$

and $I_0(x)$ denotes the modified Bessel function of the first kind.

Proof See Appendix B. □

Theorem 4.3 *Under the PS-PSC strategy, the power scaling laws of the two users are given by*

$$P_{primary} = P_r^R = M^2 \cdot (\mu_h^R |g| \beta^R)^2 + M \cdot \{\sigma_h^R |g|^2 (\beta^R)^2 + 2\mu_h^R |g| \beta^R \mathbb{E}[|h_d^R|]\} + \mathbb{E}[|h_d^R|^2], \tag{4.46}$$

and

$$P_{secondary} = P_r^T = \frac{4}{\pi^2} M^2 \cdot (\mu_h^T |g| \beta^T)^2 + M \cdot \{\sigma_h^R |g|^2 (\beta^T)^2 + 2\mu_h^T |g| \beta^T \mathbb{E}[|h_d^T|]\} + \mathbb{E}[|h_d^T|^2]. \tag{4.47}$$

Proof *For the primary user, the received power can be calculated as follows:*

$$P_r^R = \mathbb{E}[|H^R|^2] = \mathrm{Var}[|H^R|] + (\mathbb{E}[|H^R|])^2. \tag{4.48}$$

Thus, the power scaling law of the primary user can be obtained by substituting (4.42) and (4.43) into (4.48). For the secondary user, we have $P_r(T) = \int_0^\infty x^2 f_{|h_s^T|}(x)dx + \mathbb{E}[|h_d^T|^2]$. *By using the PDF of* $|h_s^T|$ *in (4.44), the scaling law in (4.47) is proved.* □

Remark 4.3 The power scaling law in (4.46) can be further simplified for STARSs with sufficiently large numbers of elements. For the case where $M \gg 1$, the received power simply scales with M^2 in the large M regime, i.e.,

$$P_r^\chi \propto \tilde{f}_\chi \cdot M^2, \tag{4.49}$$

where $\tilde{f}_R = (\mu_h^R \beta^R)^2$ and $\tilde{f}_T = (\frac{2}{\pi} \mu_h^T \beta^T)^2$. This result is consistent with the existing power scaling analysis for conventional reflecting-only RIS (Basar et al., 2019). However, the received power of the secondary user is reduced by a factor of $(2/\pi)^2$ compared with that of the primary user.

4.4.3 DP-PSC Strategy

In this subsection, we demonstrate that the DP-PSC strategy indeed achieves full diversity order for the users on both sides of the STARS. For the convenience of the subsequent analysis, we rewrite the phase-shift value of the m-th element for user χ as follows:

$$\theta_m^\chi = \Delta\theta_m^\chi + \delta_m^\chi, \tag{4.50}$$

where $\Delta\theta_m^\chi$ is the phase shift that achieves cophasing for user χ and $\delta_m^\chi = \theta_m^\chi - (\angle h_d^\chi - \angle h_m^\chi - \angle g_m) \mod 2\pi$ is the deviation from this optimal value due to the phase correlation. Based on the channel model presented in Sect. 4.3, by plugging (4.50) into (4.15), the overall channel for user χ can be rewritten as follows:

$$H^\chi = \beta^\chi \sum_{m=1}^{M} |g_m||h_m^\chi| \exp\{j(\angle h_d^\chi + \delta_m^\chi)\} + h_d^\chi. \tag{4.51}$$

4.4.3.1 Outage Probability

Due to the complexity of the DP-PSC strategy, it is very challenging to obtain the exact channel distribution for both users. To address this issue, we introduce the following theorem based on which the diversity orders of user T and user R can be obtained.

Theorem 4.4 *The outage probabilities of user T and user R are upper-bounded by*

$$P_{out}^\chi(\tau_\chi) \le \left(F(\tau_\chi, K_g, K_h^\chi, \Omega_g, \Omega_h^\chi)\right)^M \cdot \left(1 - Q\left(\sqrt{2K_d^\chi}, \sqrt{2\tau_\chi(K_d^\chi+1)/\Omega_d^\chi P_{BS}}\right)\right), \tag{4.52}$$

where $\tau_\chi = \tau_\chi^{NOMA}$ for NOMA and τ_χ^{OMA} for OMA, and $Q(a, b)$ is the Marcum Q function, which is defined as follows:

$$Q(a, b) = \int_b^\infty x \cdot \exp\left(-\frac{x^2 + a^2}{2}\right) I_0(ax)dx, \tag{4.53}$$

where $F(\tau_\chi, K_g, K_h^\chi, \Omega_g, \Omega_h^\chi) = \Pr\left\{|g_m||h_m^\chi| \le \sqrt{\frac{\tau_\chi}{P_{BS}}}\right\}$. A closed-form expression for F can be obtained using the exact PDF of $|g_m||h_m^\chi|$, i.e., a product of two Rician variables (Simon, 2002), which is omitted here for brevity.

Proof In (4.51), the phase difference between any two terms in the summation is always less than π, i.e., $|\delta_p^\chi - \delta_q^\chi| \le \pi, \ \forall p, q \in \{1, 2, \cdots, M\}$. Thus, according to the law of cosines, we have the following proposition:

$$|H^\chi| < \sqrt{\frac{\tau_\chi}{P_{BS}}} \Rightarrow \left(|h_d^\chi| < \sqrt{\frac{\tau_\chi}{P_{BS}}}\right) \wedge \left(|g_1||h_1^\chi| < \sqrt{\frac{\tau_\chi}{P_{BS}}}\right) \wedge \cdots \wedge \left(|g_M||h_M^\chi| < \sqrt{\frac{\tau_\chi}{P_{BS}}}\right), \tag{4.54}$$

where P $\Rightarrow$ Q means that proposition P leads to Q and $\wedge$ is the *logical AND operator*. Since $|H_1|, |H_2|, \cdots, |H_M|$ are i.i.d. random variables, according to probability theory, we have the following result:

$$P_{out}^{\chi} \leq \left(\Pr\{|g_m||h_m^{\chi}| \leq \sqrt{\frac{\tau_\chi}{P_{BS}}}\}\right)^M \cdot \Pr\{|h_d^{\chi}| \leq \sqrt{\frac{\tau_\chi}{P_{BS}}}\}. \tag{4.55}$$

Finally, the second term on the right-hand side (RHS) of (4.55) can be calculated using the Rician PDF of $|h_d^{\chi}|$ as follows:

$$\Pr\{|h_d^{\chi}| \leq \sqrt{\frac{\tau_\chi}{P_{BS}}}\} = 1 - Q(\sqrt{2K_d^{\chi}}, \sqrt{\frac{2\tau_\chi(K_d^{\chi}+1)}{\Omega_d^{\chi} P_{BS}}}). \tag{4.56}$$

This completes the proof. □

Corollary 4.2 *For the DP-PSC strategy, full diversity order is achieved for both user R and user T, i.e.,*

$$d^R = d^T = M + 1. \tag{4.57}$$

Proof The RHS of (4.55) is a multiplication of $M+1$ terms. According to the system model, $|h_d^{\chi}|$ is Rician distributed with shape parameter K_d and scale parameter $\Omega_d = \mathbb{E}[|h_d^{\chi}|^2]$. Since in the high transmit power regime, i.e., $P_{BS} \to \infty$, the outage probability is determined by the PDF of $|h_d^{\chi}|$ near zero, we consider the Taylor expansion of the PDF near the origin: $f_{|h_d^{\chi}|}(x) = \frac{2(K_d^{\chi}+1)}{\Omega_d} x + o(x)$. Therefore, we have the following result:

$$\Pr\{|h_d^{\chi}| \leq \sqrt{\frac{\tau_\chi}{P_{BS}}}\} = \int_0^{\sqrt{\frac{\tau_\chi}{P_{BS}}}} f_{|h_d^{\chi}|}(x)dx \approx \frac{2}{\Omega_h}(K_d^{\chi}+1)\frac{\tau_\chi}{P_{BS}}. \tag{4.58}$$

The remaining terms in (4.55), i.e., $\Pr\{|g_m||h_m^{\chi}| \leq \sqrt{\frac{\tau_\chi}{P_{BS}}}\}$, can be evaluated in a similar manner. Thus, we obtain for $P_{out}^{\chi}(\tau_\chi)$ the following expression:

$$P_{out}^{\chi}(\tau_\chi) \sim (P_{BS})^{-(M+1)}. \tag{4.59}$$

Based on the definition of the diversity order given in (4.35), we conclude that the full diversity order $M+1$ is achieved for both users. □

4.4.3.2 Power Scaling Law

As can be observed from Fig. 4.2b, the DP-PSC strategy is symmetric for the two users. As a result, similar power scaling laws apply to both users. For simplicity, we consider the scenario, where M is large and the power carried by the direct channel is negligible. For the end-to-end channel given in (4.51), we obtain the following theorem.

Theorem 4.5 *For the DP-PSC strategy, the power scaling laws for both users can be expressed as follows:*

$$P_r^\chi \propto \frac{8}{\pi^2}M^2(\mu_h^\chi|g|\beta^\chi)^2 + \left(1-\frac{2}{\pi}\right)M\sigma_h^\chi|g|^2(\beta^\chi)^2. \tag{4.60}$$

Proof The received power can be expressed as $P_r^\chi = \mathbb{E}[|H^\chi|^2]$. Furthermore, by substituting $\Delta^\chi = \pi/2$ in the proof process given in Appendix B, the approximate PDF of $|H^\chi|$ is obtained as

$$f_{|H^\chi|}(x) = \frac{x}{\beta^2}e^{-\frac{x^2+\alpha^2}{2\beta^2}}I_0\left(\frac{x\alpha}{\beta}\right), \tag{4.61}$$

where

$$\alpha = 2\sqrt{2}M\mu_h^\chi/\pi,\ \beta^2 = (1-2/\pi)\frac{M}{2}\Omega_h^\chi. \tag{4.62}$$

Finally, by integrating the PDF obtained in (4.61), the proof is completed. □

Remark 4.4 Similar to the PS-PSC strategy, the average received power under the DP-PSC strategy also scales with M^2 in the large M regime. However, under the same channel conditions and using equal power splitting, i.e., $\beta^R = \beta^T$, comparing the average power received by the primary user and secondary user for the PS-PSC strategy, $P_{\text{PS-PSC}}^{\text{primary}}$ and $P_{\text{PS-PSC}}^{\text{secondary}}$, respectively, and the average power received by the users for the DP-PSC strategy, $P_{\text{DP-PSC}}$, we have

$$P_{\text{PS-PSC}}^{\text{primary}} > P_{\text{DP-PSC}} > P_{\text{PS-PSC}}^{\text{secondary}}. \tag{4.63}$$

4.4.4 TR-PSC Strategy

For the TR-PSC strategy, the STARS is equivalent to a composite smart surface containing a reflecting-only RIS with M_R elements and a transmitting-only RIS with M_T elements, where $M_R + M_T = M$. As a result, for each reflecting-only/transmitting-only element, the phase shift of the RC/TC can be configured without taking into account the correlation in (2.11). As cophasing can be achieved on both sides, the end-to-end channel for user χ is given by

$$H^\chi = \sum_{m'=1}^{M_\chi}|g_{m'}||h_{m'}^\chi| + h_d^\chi. \tag{4.64}$$

4.4.4.1 Outage Probability

Considering (4.64), the end-to-end channel is the sum of the amplitudes of the cascaded channels $|g_{m'}||h^{\chi}_{m'}|$, which are perfectly aligned in phase. Thus, this expression is in a similar form as the end-to-end channel of the primary user for the PS-PSC strategy.

Theorem 4.6 *For the TR-PSC strategy, the asymptotic outage probability for both users can be expressed as follows:*

$$P^{\chi}_{out}(\tau_\chi) = \frac{2^{M_\chi+1}(K^R_h+1)^{M_\chi}(K^R_d+1)}{(2M_\chi+2)!\,(\Omega^R_h)^{M_\chi}\Omega^R_d}(\beta^R)^{-M_\chi/2} \cdot e^{-M_\chi K_s - K_d}(\tau_\chi)^{M_\chi+1}P_{BS}^{-(M_\chi+1)}, \tag{4.65}$$

where $\tau_\chi = \tau^{NOMA}_\chi$ *for NOMA and* $\tau_\chi = \tau^{OMA}_\chi$ *for OMA.*

Proof The proof of this theorem is similar to that of Theorem 4.2 and is omitted here for brevity. □

According to (4.65), for the TR-PSC strategy, the diversity orders are given by

$$d^R = M_R + 1 \text{ and } d^T = M_T + 1. \tag{4.66}$$

4.4.4.2 Power Scaling Law

By comparing the PSC of the TR-PSC strategy in (4.21) and (4.22) with the configuration in (4.18), we observe that the power scaling laws of the TR-PSC strategy can be deduced from Theorem 4.3.

Corollary 4.3 *For the TR-PSC strategy, the power scaling law for both users can be expressed as follows:*

$$P^{\chi}_r \propto M^2_\chi \cdot (\mu^{\chi}_h|g|\beta^\chi)^2 + M_\chi \cdot \left\{\sigma^{\chi}_h|g|^2(\beta^\chi)^2 + 2\mu^{\chi}_h|g|\beta^\chi\mathbb{E}[|h^R_d|]\right\} + \mathbb{E}[|h^R_d|^2]. \tag{4.67}$$

Proof The proof of this corollary is similar to the proof of Theorem 4.3 and is omitted for brevity. □

Remark 4.5 Corollary 4.3 provides an important insight regarding the differences between the power scaling laws for the DP-PSC and TR-PSC strategies. For a STARS with a total of M elements, in the large M regime, the received power for DP-PSC scales with M^2 on both sides of the surface, while TR-PSC can only produce powers that scale with respectively M^2_R and M^2_T on the two sides of the STARS. Furthermore, since $M_R, M_T \leq M$, the power differences between the two strategies are more pronounced for large numbers of elements.

4.4.5 Performance Lower and Upper Bounds

To have performance baselines for the proposed PSC strategies for STARS-aided communication systems, we proposed lower and upper bounds. For the performance lower bound, we assume that the STARS employs a random PSC strategy. For the performance upper bound, the STARS is assumed to be able to perform independent phase-shift adjustments for both the transmitted and reflected signals. The end-to-end channel gain and the achievable rate of the primary user for the PS-PSC strategy achieve this performance upper bound. This is because for the primary user, the STARS phase-shift coefficients are chosen without considering the phase correlation constraint. Thus, by designating user R or user T as the primary user in (4.36) and (4.46), the diversity order and power scaling law of the proposed performance upper bound are achieved. Consequently, in the following, we focus on the analysis of the proposed performance lower bound.

4.4.5.1 Outage Probability

For the random PSC strategy, we assume that for all m, the phase shifts of the RCs are randomly chosen, i.e., θ_m^R is uniformly distributed within $[0, 2\pi)$. The overall end-to-end channel between the BS and user χ is given by

$$H^\chi = |g|\beta^\chi \sum_{m=1}^{M} |h_m^\chi| \exp\{j(\angle g_m + \angle h_m^\chi + \theta_m^\chi)\} + h_d^\chi. \tag{4.68}$$

Moreover, due to the randomness of the STARS phase shifts, θ_m^χ, the entire phase term in (4.68) can be regarded as a uniformly distributed random variable, which leads to the following theorem. For convenience, we denote the first term in (4.68) as h_s^χ, and thus the overall channel is given by $H^\chi = h_s^\chi + h_d^\chi$.

Theorem 4.7 *For the random PSC strategy, if the number of STARS elements is large,* $|h_s^\chi|$ *can be approximated by the following PDF:*

$$f_{|h_s^\chi|}(x) = \frac{2x}{\Omega_r^\chi} e^{-\frac{x^2}{\Omega_r^\chi}}, \tag{4.69}$$

where $\Omega_r^\chi = M|g|^2(\beta^\chi)^2\Omega_h^\chi$.

Proof *We consider the real (in-phase) and imaginary (quadrature) components of* h_s^χ*:*

$$T_c = |g|\beta^\chi \sum |h_m^\chi| \cos(\angle g_m + \angle h_m^\chi + \theta_m^\chi), \tag{4.70}$$

$$T_s = |g|\beta^\chi \sum |h_m^\chi| \sin(\angle g_m + \angle h_m^\chi + \theta_m^\chi). \tag{4.71}$$

According to the central limit theorem (CLT), for large M, the distributions of T_c and T_s can be approximated by Gaussian distributions with zero means and variances $\mathbb{E}[T_c^2] = \mathbb{E}[T_s^2] = M|g|^2(\beta^\chi)^2\Omega_h^\chi/2$. Thus, $|h_s^\chi| = \sqrt{T_c^2 + T_s^2}$ follows a Rayleigh distribution with the PDF given in (4.69). □

For the considered system, the outage probability of user χ can be analyzed as follows:

$$
\begin{aligned}
P_{out}^\chi(\gamma_0^\chi) &= \Pr\left\{\frac{(|h_s^\chi|^2 + |h_d^\chi|^2)c_\chi^2 P_{BS}}{\sigma_0^2} < \gamma_0^\chi\right\} \\
&\leq \Pr\left\{|h_s^\chi|^2 < \frac{\tau_\chi}{P_{BS}}\right\} \cdot \Pr\left\{|h_d^\chi|^2 < \frac{\tau_\chi}{P_{BS}}\right\} \\
&= \left(1 - e^{-\frac{\tau_\chi}{\Omega_r^\chi P_{BS}}}\right) \cdot \left(1 - Q(\sqrt{2K_d^\chi}, \sqrt{\frac{2\tau_\chi(K_d^\chi+1)}{\Omega_d^\chi P_{BS}}})\right).
\end{aligned}
\tag{4.72}
$$

Thus, for the random PSC strategy, the diversity order of both users can be calculated using (4.35) and (4.72):

$$
d_{\text{random}}^R = d_{\text{random}}^T = 2. \tag{4.73}
$$

4.4.5.2 Power Scaling Law

Considering the PDF of h_s^χ given in (4.69), we have $\mathbb{E}[|h_s^\chi|] = \Omega_r^\chi = M|g|^2(\beta^\chi)^2\Omega_h^\chi$. Then, the power scaling law for the random PSC strategy is given by

$$
P_r^\chi \propto \mathbb{E}[|h_s^\chi|^2] + \mathbb{E}[|h_d^\chi|^2] = M|g|^2(\beta^\chi)^2\Omega_h^\chi + \mathbb{E}[|h_d^\chi|^2]. \tag{4.74}
$$

Note that the result in (4.74) scales linearly with M. Thus, compared to the three proposed PSC strategies, the average power received for the users increases significantly slower with M for the random PSC strategy.

4.4.6 Summary and Discussion

The main results of this section are summarized in Table 4.2.[4] It can be observed that the PS-PSC strategy is suitable for creating asymmetric channel gains, while the DP-PSC strategy is suitable for balancing the performances of both users. Moreover,

[4] In this table, user R is assumed to be the primary user for the PS-PSC strategy.

Table 4.2 Comparing different PSC strategies

STARS phase-shift strategy	PS-PSC	DP-PSC	TR-PSC	Random PSC
Diversity order of user R	$M+1$	$M+1$	M_R+1	2
Diversity order of user T	$(M+3)/2$	$M+1$	M_T+1	2
Power scaling law	$\propto M^2$	$\propto M^2$	$\propto M_\chi^2$	$\propto M$

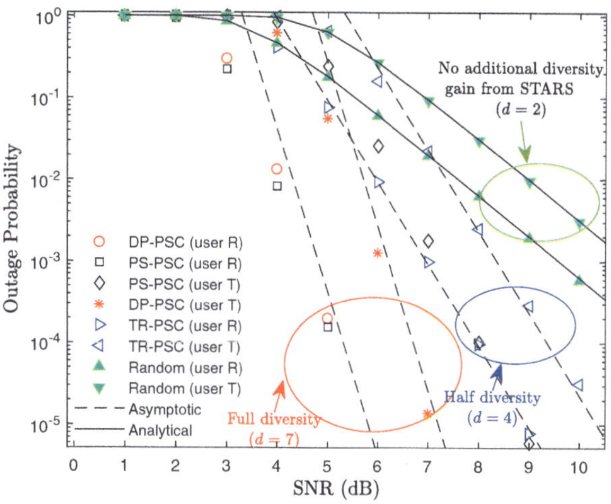

Fig. 4.3 Outage probabilities for OMA users employing different PSC strategies

Table 4.2 reveals that the DP-PSC strategy provides the best performance in terms of diversity gain and power scaling.

4.4.7 *Numerical Results*

For this simulation, we assume all STARS-user channels are modeled as Rician fading channels with path loss exponent $\alpha = 2.2$ and Rician factor $K = 1.3$ dB. The noise power for both users is set to $\sigma_0^2 = -50$ dBm. We also assume that the scale factor for user R is twice as the scale factor of user T. Thus, the average received power of user R is larger. The transmit SNR (P_{BS}/σ_0^2) was varied between 0 dB and 12 dB. We consider an asymmetric channel by setting the expected power of the T&R channels as $\mathbb{E}[|h_m^R|^2] = 2\mathbb{E}[|h_m^T|^2]$. The target data rate for both users is set at $\tilde{R} = 1.5$ bit/s/Hz. In Fig. 4.3, we present simulation results for the outage probabilities of both users for the proposed PSC strategies and the baseline random PSC strategy. To verify our performance analysis, asymptotic results are presented for the proposed PSC strategies exploiting (4.34) and (4.65), while analytical results are presented for the random PSC strategy exploiting (4.72).

As can be observed, the primary user for the PS-PSC strategy achieves the lowest outage probability and full diversity order, since the STARS can align the reflected

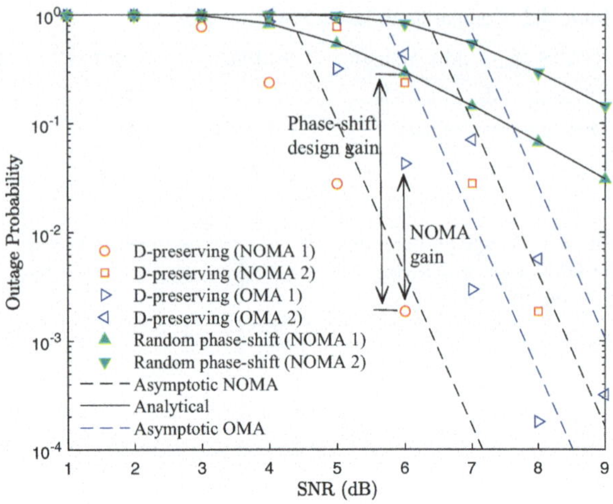

Fig. 4.4 Outage probabilities for NOMA and OMA users for the DP-PSC strategy

channels with the direct channel for the primary user. However, due to the coupled phase-shift model, the secondary user suffers a noticeable performance loss as the available choices for the transmission phase shifts are limited. Moreover, it can be observed that the primary user and the secondary user achieve full diversity order and reduced diversity order, respectively, which are consistent with (4.36) and (4.38). For the DP-PSC strategy, as can be seen from Fig. 4.3, the outage probabilities of both users exhibit similar slopes and can achieve full diversity order, which is consistent with (4.57). In addition, user R still has a lower outage probability than user T. This is because user R has a higher channel gain due to its shorter path length. For the TR-PSC strategy, since only half of the STARS elements are used for transmission/reflection (i.e., reduced array gain), there is a considerable performance loss for both users and only half of the maximum diversity order can be achieved, as is expected from (4.66). Furthermore, the random PSC strategy yields the worst performance and achieves no additional diversity gain, which is in accordance with (4.73).

In Fig. 4.4, we compare the outage probabilities of the users for NOMA and OMA. The STARS is assumed to employ the DP-PSC strategy. The power allocation factors for the two users are set to $c_R^2 = 0.4$ and $c_T^2 = 0.6$. As can be observed, for both users, NOMA yields a better outage performance than OMA. Moreover, comparing the DP-PSC strategy with the random PSC strategy, it can be observed that the performance gap between the two PSC strategies (illustrated with the vertical double arrow) increases with the transmit SNR. In contrast, the performance gain of NOMA over OMA in terms of outage probability stays constant in the high SNR regime, which is consistent with (4.52).

4.5 Summary

In this chapter, we studied the communication performance of STARS under independent and coupled phase-shift models. For the independent phase-shift model, we analyzed the diversity gain that can be achieved by the STARS on both sides. For the coupled phase-shift model, we first proposed three practical PSC strategies. We then derived the outage probability and power scaling law that can be achieved by each strategy, which was further demonstrated by numerical results.

References

Basar, E., Di Renzo, M., De Rosny, J., Debbah, M., Alouini, M.-S., & Zhang, R. (2019). Wireless communications through reconfigurable intelligent surfaces. *IEEE Access, 7*, 116753–116773.

Liu, Y., Liu, X., Mu, X., Hou, T., Xu, J., Di Renzo, M., & Al-Dhahir, N. (2021). Reconfigurable intelligent surfaces: Principles and opportunities. *IEEE Communications Surveys & Tutorials, 23*, 1546–1577.

Liu, Y., Qin, Z., Elkashlan, M., Gao, Y., & Hanzo, L. (2017). Enhancing the physical layer security of non-orthogonal multiple access in large-scale networks. *IEEE Transactions on Wireless Communications, 16*, 1656–1672.

Marzetta, T. L., & Ngo, H. Q. (2016). *Fundamentals of massive MIMO*. Cambridge University Press.

Simon, M. K. (2002). *Probability distributions involving Gaussian random variables: A handbook for engineers and scientists*. Springer.

Wang, T., Chen, G., Coon, J. P., & Badiu, M.-A. (2020). Study of intelligent reflective surface assisted communications with one-bit phase adjustments. In *Proceedings of the IEEE Global Communication Conferemce (GLOBECOM)* (pp. 1–6).

Chapter 5
Operating Protocols and Beamforming Design for STARS

Abstract In this chapter, we will introduce the operating protocols for employing STARS in wireless communications and discuss the corresponding beamforming design problems. Both independent and coupled phase-shift models will be considered. Moreover, a unified optimization framework will be introduced for STARS beamforming design.

5.1 Operating Protocols for STARS

As previously introduced, the wireless signal incident on a given element of the STARS is divided into a transmitted and a reflected signal. To characterize this feature, let s_n denote the signal incident on the n-th element of the STARS, where $n \in \mathcal{N} \triangleq \{1, 2, \ldots, N\}$ and N denotes the total number of elements. The signals transmitted and reflected by the n-th element can be modeled as $t_n = \left(\sqrt{\beta_n^t} e^{j\theta_n^t}\right) s_n$ and $r_n = \left(\sqrt{\beta_n^r} e^{j\theta_n^r}\right) s_n$, respectively, where $\sqrt{\beta_n^t} \in [0, 1], \theta_n^t \in [0, 2\pi)$ and $\sqrt{\beta_n^r} \in [0, 1], \theta_n^r \in [0, 2\pi)$ denote the amplitude and phase-shift response of the n-th element's TRCs. It can be observed that by properly adjusting the amplitude coefficients for transmission and reflection, a given STARS element can be operated in the full transmission mode (i.e., $\beta_n^t = 1, \beta_n^r = 0$, referred to as T mode), the full reflection mode (i.e., $\beta_n^t = 0, \beta_n^r = 1$, referred to as R mode), and the general simultaneous transmission and reflection mode ($\beta_n^t, \beta_n^r \in [0, 1]$, referred to as T&R mode). Inspired by these observations, we propose three practical protocols for operating STARS in wireless communication systems, namely energy splitting (ES), mode switching (MS), and time switching (TS), as illustrated in Fig. 5.1.

5.1.1 Energy Splitting

For ES, as shown in Fig. 5.1a, all elements of the STARS are assumed to operate in the T&R mode, where the energy of the signal incident on each element is generally split into the energies of the transmitted

Y. Liu et al., *Simultaneously Transmitting and Reflecting Surfaces for Wireless Communications*, Wireless Networks, https://doi.org/10.1007/978-3-031-76848-4_5

and reflected signals with an energy-splitting ratio of $\beta_n^t : \beta_n^r$. In this case, the transmission- and reflection-coefficient matrices of the STARS are given by $\mathbf{\Theta}_t^{\mathrm{ES}} = \mathrm{diag}\left(\sqrt{\beta_1^t}e^{j\theta_1^t}, \sqrt{\beta_2^t}e^{j\theta_2^t}, \ldots, \sqrt{\beta_N^t}e^{j\theta_N^t}\right)$ and $\mathbf{\Theta}_r^{\mathrm{ES}} = \mathrm{diag}\left(\sqrt{\beta_1^r}e^{j\theta_1^r}, \sqrt{\beta_2^r}e^{j\theta_2^r}, \ldots, \sqrt{\beta_N^r}e^{j\theta_N^r}\right)$, respectively, where $\beta_n^t, \beta_n^r \in [0, 1]$, $\beta_n^t + \beta_n^r = 1$, and $\theta_n^t, \theta_n^r \in [0, 2\pi), \forall n \in \mathcal{N}$. For ES, since both the TRCs of each element can be optimized, a high degree of flexibility for communication system design is enabled. However, the large number of design variables also causes a relatively high overhead for configuration information exchange between the BS and the STARS.

5.1.2 Mode Switching

For MS, as shown in Fig. 5.1b, all elements of the STARS are divided into two groups. Specifically, one group contains N^t elements that operate in the T mode,

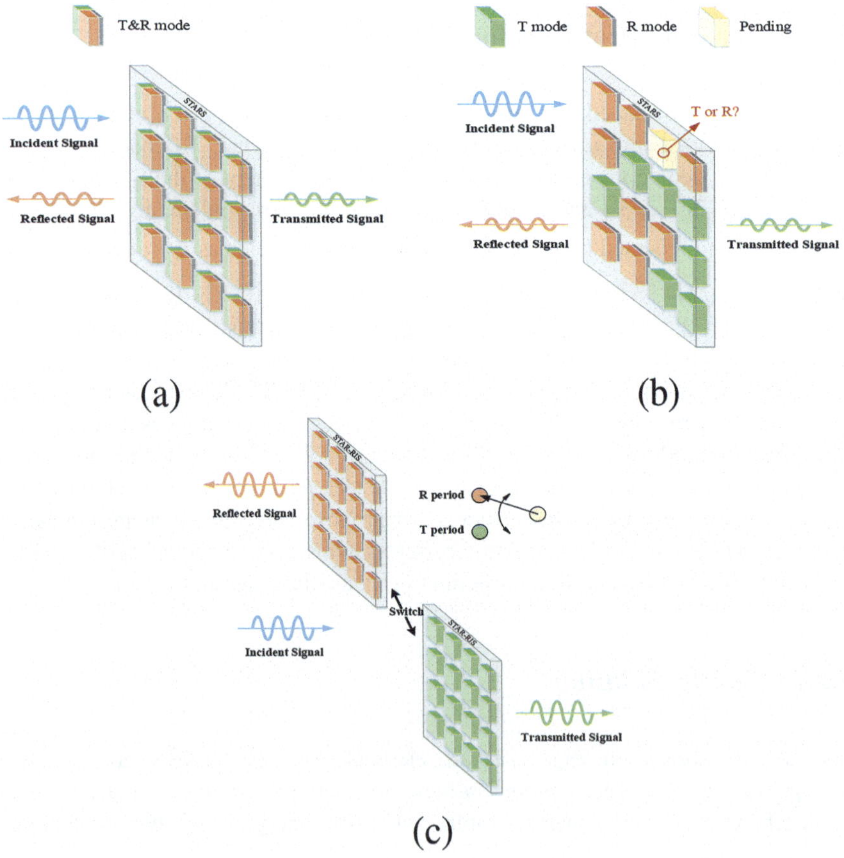

Fig. 5.1 Illustration of three practical protocols for operating STARS. (**a**) Energy splitting (ES). (**b**) Mode switching (MS). (**c**) Time switching (TS)

while the other group contains N^r elements operating in the R mode, where $N^t + N^r = N$. Accordingly, the STARS transmission- and reflection-coefficient matrices are given by $\mathbf{\Theta}_t^{\text{MS}} = \text{diag}\left(\sqrt{\beta_1^t}e^{j\theta_1^t}, \sqrt{\beta_2^t}e^{j\theta_2^t}, \ldots, \sqrt{\beta_N^t}e^{j\theta_N^t}\right)$ and $\mathbf{\Theta}_r^{\text{MS}} = \text{diag}\left(\sqrt{\beta_1^r}e^{j\theta_1^r}, \sqrt{\beta_2^r}e^{j\theta_2^r}, \ldots, \sqrt{\beta_N^r}e^{j\theta_N^r}\right)$, respectively, where $\beta_n^t, \beta_n^r \in \{0, 1\}$, $\beta_n^t + \beta_n^r = 1$, and $\theta_n^t, \theta_n^r \in [0, 2\pi), \forall n \in \mathcal{N}$. We note that MS STARS can be regarded as a special case of ES STARS, where the amplitude coefficients for transmission and reflection are restricted to binary values. Therefore, MS generally cannot achieve the same full-dimension transmission and reflection beamforming gain as ES, since only a subset of the elements are selected for transmission and reflection, respectively. Nevertheless, MS is still appealing in practice, since such an "on-off" type operating protocol is much easier to implement compared to the ES protocol.

5.1.3 Time Switching

Different from ES and MS, the TS STARS exploits the time domain and periodically switches all elements between the T mode and the R mode in different orthogonal time slots (referred to as T and R period), as illustrated in Fig. 5.1c. Let $0 \leq \lambda^t \leq 1$ and $0 \leq \lambda^r \leq 1$ denote the percentage of communication time allocated to the T period and R period, respectively, where $\lambda^t + \lambda^r = 1$. Therefore, the corresponding STARS transmission- and reflection-coefficient matrices are given by $\mathbf{\Theta}_t^{\text{TS}} = \text{diag}\left(e^{j\theta_1^t}, e^{j\theta_2^t}, \ldots, e^{j\theta_N^t}\right)$ and $\mathbf{\Theta}_r^{\text{TS}} = \text{diag}\left(e^{j\theta_1^r}, e^{j\theta_2^r}, \ldots, e^{j\theta_N^r}\right)$, where $\theta_n^t, \theta_n^r \in [0, 2\pi), \forall n \in \mathcal{N}$. Different from the ES and MS protocols, due to the exploitation of the time domain, the design of the TRCs for TS is not coupled and thus easier to handle. However, the periodical switching of the elements introduces stringent requirements for time synchronization, which entails a higher hardware implementation complexity.

In Table 5.1, we summarize the unique ES, MS, and TS optimization variables and identify their respective advantages and disadvantages. It is worth mentioning that the aforementioned coupled phase-shift model in (2.12) is only valid and has to be considered in ES STARS. As a result, the transmission and reflection beamforming design associated with ES STARS is more challenging than MS STARS and TS STARS. In the following, we will discuss the STARS beamforming design with a particular focus on the ES protocol.

5.2 STARS Beamforming Design Under the Independent Phase-Shift Model

Compared to conventional reflecting/transmitting-only RISs, the STARS beamforming design imposes unique challenges. Firstly, there are significantly increased numbers of optimization variables, i.e., the optimization of amplitudes and phase

Table 5.1 Summary of Proposed Protocols for Operating STARS

Protocols	Optimization variables	Advantages	Disadvantages
ES	• Amplitude and phase-shift coefficients of each element for transmission and reflection	High flexibility	Large number of design variables
MS	• Mode selection of each element • Transmission phase-shift coefficients for T mode elements • Reflection phase-shift coefficients for R mode elements	Easy to implement	Reduced transmission and reflection gain
TS	• Time allocation • Transmission phase-shift coefficients of each element during T period • Reflection phase-shift coefficients of each element during R period	Independent T and R designs	High hardware implementation complexity

shifts for both transmission and reflection, which further increases the design complexity. Secondly, additional coupling constraints exist in the STARS beamforming configuration (e.g., the energy conservation constraint, the coupled phase-shift model, etc.), which make the design quite challenging. In this section, we focus our attention on the ES STARS beamforming design under the independent phase-shift model, where an iterative joint optimization algorithm is proposed.

5.2.1 System Model and Problem Formulation

Focusing on ES STARS, we present the system model of a STARS-aided downlink communication system and formulate joint active and passive beamforming optimization problems for both unicast and multicast transmission.

5.2.1.1 System Model

As shown in Fig. 5.2, we consider a narrowband STARS-aided downlink communication system operating over frequency-flat channels, where an N-antenna BS communicates with multiple single-antenna users with the aid of a STARS comprising N elements. As previously discussed, the STARS can create full-space coverage by simultaneously transmitting and reflecting the incident signal. We refer to users that are located in the transmission space as T users, while users that are located in the reflection space are referred to as R users. As depicted in Fig. 5.2, in this section, we assume that the direct communication links between the BS and the users are blocked by obstacles, as this is one of the most challenging

Fig. 5.2 Illustration of a STARS-aided downlink communication system, where the direct BS-user links are blocked by obstacles

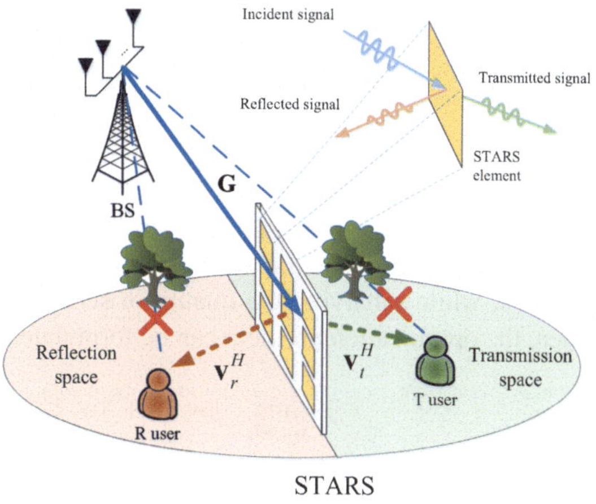

scenarios for conventional communication systems. The STARS is deployed to provide communication service for the users in the resulting signal dead zone by establishing additional transmission and reflection links. Here, we consider a two-user setup, i.e., the STARS-aided communication system serves one T user and one R user. Let $\mathbf{G} \in \mathbb{C}^{N\times M}$, $\mathbf{v}_t^H \in \mathbb{C}^{1\times N}$, and $\mathbf{v}_r^H \in \mathbb{C}^{1\times N}$ represent the narrowband quasi-static fading channels from the BS to the STARS, the STARS to T user, and the STARS to R user, respectively. In order to be able to unveil the maximum performance gain enabled by STARS, the perfect CSI of all channels is assumed to be available at the BS.

In this section, we study the joint design of the active beamforming at the BS and the passive transmission and reflection beamforming at the STARS for both unicast and multicast communication. For unicast, the BS sends independent information to the T and R users. For multicast, the BS sends the same information to both users.

5.2.1.2 Unicast Communication and Problem Formulation

In unicast transmission, let $\mathbf{w}_k$ and x_k denote the active beamforming vector and the information-bearing symbol for user $k \in \{t, r\}$ at the BS, respectively. For ES STARS, let $\mathbf{\Theta}_k = \mathrm{diag}\left(\sqrt{\beta_1^k}e^{j\theta_1^k}, \sqrt{\beta_2^k}e^{j\theta_2^k}, \ldots, \sqrt{\beta_N^k}e^{j\theta_N^k}\right), \forall k \in \{t, r\}$ denote the transmission/reflection-coefficient matrix of the ES STARS. The received signal at user $k \in \{t, r\}$ is given by

$$y_{\mathrm{UC},k} = \mathbf{v}_k^H \mathbf{\Theta}_k \mathbf{G} \left(\mathbf{w}_t x_t + \mathbf{w}_r x_r\right) + n_k, \tag{5.1}$$

where $\mathbb{E}\left[|x_k|^2\right] = 1$ and $n_k \sim \mathcal{CN}\left(0, \sigma_k^2\right)$ denotes the AWGN at user k. Therefore, the achievable communication rate of user $k \in \{t, r\}$ is given by

$$R_{\mathrm{UC},k} = \log_2 \left(1 + \frac{\left| \mathbf{v}_k^H \mathbf{\Theta}_k \mathbf{G} \mathbf{w}_k \right|^2}{\left| \mathbf{v}_k^H \mathbf{\Theta}_k \mathbf{G} \mathbf{w}_{\overline{k}} \right|^2 + \sigma_k^2} \right), \tag{5.2}$$

where $\overline{k} = r$, if $k = t$; and $\overline{k} = t$, otherwise.

For unicast transmission, we aim to minimize the total power consumption of the BS by jointly optimizing the active beamforming at the BS and the passive transmission and reflection beamforming at the STARS for a given operating protocol, while satisfying the quality-of-service (QoS) requirement of both users. Then, the optimization problem can be formulated as follows:

$$\min_{\mathbf{w}_k, \mathbf{\Theta}_k} \quad \|\mathbf{w}_t\|^2 + \|\mathbf{w}_r\|^2 \tag{5.3a}$$

$$\text{s.t.} \quad R_{\mathrm{UC},k} \geq \overline{R}_k, \forall k \in \{t, r\}, \tag{5.3b}$$

$$\theta_n^t, \theta_n^r \in [0, 2\pi), \forall n \in \mathcal{N}, \tag{5.3c}$$

$$0 \leq \beta_n^t, \beta_n^r \leq 1, \beta_n^t + \beta_n^r = 1, \forall n \in \mathcal{N}, \tag{5.3d}$$

where $\overline{R}_k$ denotes the minimum rate requirement of user k. Let $\mathbf{H} = [\mathbf{h}_t \ \mathbf{h}_r] \in \mathbb{C}^{M \times 2}$, where $\mathbf{h}_k^H = \mathbf{v}_k^H \mathbf{\Theta}_k \mathbf{G}$, $\forall k \in \{t, r\}$. Assuming that $N \geq 2$, a sufficient condition for the feasibility of problem (5.3) for any finite $\overline{R}_k$ is that Rank $(\mathbf{H}) = 2$, where a zero-forcing solution can be easily constructed for the active beamforming (Wu & Zhang, 2019). Given the independent STARS-user channels, $\{\mathbf{v}_k^H\}$, and the different transmission- and reflection-coefficient matrices, $\{\mathbf{\Theta}_k\}$, such a rank condition can be satisfied with a high probability.

5.2.1.3 Multicast Communication and Problem Formulation

For multicast transmission, the BS employs one active beamforming vector, denoted by $\mathbf{w}_c$, to convey the same symbol, s_c, to the T and R users. As a result, the corresponding received signal at user $k \in \{t, r\}$ is given by

$$y_{\mathrm{MC},k} = \mathbf{v}_k^H \mathbf{\Theta}_k \mathbf{G} \mathbf{w}_c s_c + n_k. \tag{5.4}$$

Accordingly, the achievable communication rate of user $k \in \{t, r\}$ can be expressed as

$$R_{\mathrm{MC},k} = \log_2 \left(1 + \frac{\left| \mathbf{v}_k^H \mathbf{\Theta}_k^{\mathrm{ES/MS}} \mathbf{G} \mathbf{w}_c \right|^2}{\sigma_k^2} \right). \tag{5.5}$$

Since the information has to be delivered to both users, the system performance of multicast communication is limited by the user achieving the smaller communication rate. As a result, the effective system multicasting rate for the considered operating protocols is given by

$$R_{\mathrm{MC}} = \min \left\{ R_{\mathrm{MC},t}, R_{\mathrm{MC},r} \right\}. \tag{5.6}$$

Then, for multicast transmission, the joint beamforming optimization problem for ES STARS can be formulated as

$$\min_{\mathbf{w}_c, \mathbf{\Theta}_k} \quad \|\mathbf{w}_c\|^2 \tag{5.7a}$$

$$\text{s.t.} \quad R_{\mathrm{MC}} \geq \overline{R}_c, \tag{5.7b}$$

$$\theta_n^t, \theta_n^r \in [0, 2\pi), \forall n \in \mathcal{N}, \tag{5.7c}$$

$$0 \leq \beta_n^t, \beta_n^r \leq 1, \beta_n^t + \beta_n^r = 1, \forall n \in \mathcal{N}, \tag{5.7d}$$

where (5.7b) denotes the minimum system multicast rate constraint with required communication rate $\overline{R}_c$. As there is no inter-user interference in multicast, problem (5.7) is always feasible for any finite $\overline{R}_c$.

5.2.1.4 Discussion

Note that the formulated joint beamforming optimization problems for multicast transmission can be regarded as special cases of those for unicast transmission. More specifically, problem (5.7) for multicast transmission can be obtained from problem (5.3) for unicast transmission by setting $\mathbf{w}_t = \mathbf{w}_r = \mathbf{w}_c$, $\overline{R}_t = \overline{R}_r = \overline{R}_c$ and removing the inter-user interference term in the denominator of the unicast communication rate expression in (5.2). Motivated by this, in the following, we mainly focus on joint beamforming optimization problem (5.3) for unicast transmission since problem (5.7) for multicast transmission can be solved in a similar manner. However, problem (5.3) is a non-convex optimization problem since the left-hand side (LHS) of (5.3b) is not concave, as $\left\{\mathbf{\Theta}_k^{\mathrm{X}}\right\}$ and $\{\mathbf{w}_k\}$ are highly coupled. Therefore, it is difficult to obtain a globally optimal solution for such a challenging problem. In the following, we develop an efficient iterative joint optimization algorithm to find a high-quality suboptimal solution for the STARS beamforming design.

5.2.2 *Iterative Joint Optimization Algorithm*

To start with, we first transform problem (5.3) into a more tractable form. To facilitate the design, we define the transmission- and reflection-coefficient vectors for ES STARS as $\mathbf{q}_k = \left[\sqrt{\beta_1^k}e^{j\theta_1^k}, \sqrt{\beta_2^k}e^{j\theta_2^k}, \ldots, \sqrt{\beta_N^k}e^{j\theta_N^k}\right]^H$, $\forall k \in \{t, r\}$, which leads to $\left|\mathbf{v}_k^H \boldsymbol{\Theta}_k \mathbf{G} \mathbf{w}_k\right|^2 = \left|(\mathbf{q}_k)^H \mathbf{H}_k \mathbf{w}_k\right|^2$, where $\mathbf{H}_k = \operatorname{diag}\left(\mathbf{v}_k^H\right)\mathbf{G}$. Moreover, we define $\mathbf{Q}_k = \mathbf{q}_k(\mathbf{q}_k)^H$, $\forall k \in \{t, r\}$, which satisfies $\mathbf{Q}_k \succeq 0$, $\operatorname{Rank}(\mathbf{Q}_k) = 1$, and $\operatorname{Diag}(\mathbf{Q}_k) = \boldsymbol{\beta}^k$, where $\boldsymbol{\beta}^k \triangleq \left[\beta_1^k, \beta_2^k, \ldots, \beta_N^k\right]$. Similarly, we define $\mathbf{W}_k = \mathbf{w}_k\mathbf{w}_k^H$, $\forall k \in \{t, r\}$, which satisfies $\mathbf{W}_k \succeq 0$ and $\operatorname{Rank}(\mathbf{W}_k) = 1$. Then, optimization problem (5.3) can be reformulated as follows:

$$\min_{\mathbf{W}_k, \mathbf{Q}_k, \boldsymbol{\beta}_k} \quad \operatorname{Tr}(\mathbf{W}_t) + \operatorname{Tr}(\mathbf{W}_r) \tag{5.8a}$$

$$\text{s.t.} \;\; \overline{\gamma}_k \operatorname{Tr}\left(\mathbf{Q}_k \mathbf{H}_k \mathbf{W}_{\overline{k}} \mathbf{H}_k^H\right) - \operatorname{Tr}\left(\mathbf{Q}_k \mathbf{H}_k \mathbf{W}_k \mathbf{H}_k^H\right) + \overline{\gamma}_k \sigma_k^2 \le 0, \forall k \in \{t, r\}, \tag{5.8b}$$

$$\operatorname{Diag}(\mathbf{Q}_k) = \boldsymbol{\beta}^k, \forall k \in \{t, r\}, \tag{5.8c}$$

$$\operatorname{Rank}(\mathbf{Q}_k) = 1, \forall k \in \{t, r\}, \tag{5.8d}$$

$$\operatorname{Rank}(\mathbf{W}_k) = 1, \forall k \in \{t, r\}, \tag{5.8e}$$

$$\mathbf{Q}_k \succeq 0, \mathbf{W}_k \succeq 0, \forall k \in \{t, r\}, \tag{5.8f}$$

$$0 \le \beta_n^t, \beta_n^r \le 1, \beta_n^t + \beta_n^r = 1, \forall n \in \mathcal{N}, \tag{5.8g}$$

where $\overline{\gamma}_k = 2^{\overline{R}_k} - 1$, $\forall k \in \{t, r\}$. Problem (5.8) is still non-convex due to the highly coupled non-convex constraint (5.8b) and the non-convex rank-one constraint (5.8d) and (5.8e).

Before handling the highly coupled non-convex constraint (5.8b), we first have the following lemma.

Lemma 5.1 *For any two Hermitian matrices $\mathbf{A} \in \mathbb{H}^N$ and $\mathbf{B} \in \mathbb{H}^N$ having the same size, we have the following two equalities:*

$$\operatorname{Tr}(\mathbf{AB}) = \frac{1}{2}\|\mathbf{A} + \mathbf{B}\|_F^2 - \frac{1}{2}\|\mathbf{A}\|_F^2 - \frac{1}{2}\|\mathbf{B}\|_F^2, \tag{5.9a}$$

$$-\operatorname{Tr}(\mathbf{AB}) = \frac{1}{2}\|\mathbf{A} - \mathbf{B}\|_F^2 - \frac{1}{2}\|\mathbf{A}\|_F^2 - \frac{1}{2}\|\mathbf{B}\|_F^2. \tag{5.9b}$$

Proof We first prove Eq. (5.9a), whose RHS can be rewritten as

$$\begin{aligned}
&\frac{1}{2}\|\mathbf{A} + \mathbf{B}\|_F^2 - \frac{1}{2}\|\mathbf{A}\|_F^2 - \frac{1}{2}\|\mathbf{B}\|_F^2 \\
&= \frac{1}{2}\operatorname{Tr}\left((\mathbf{A} + \mathbf{B})^H(\mathbf{A} + \mathbf{B})\right) - \frac{1}{2}\operatorname{Tr}\left(\mathbf{A}^H\mathbf{A}\right) - \frac{1}{2}\operatorname{Tr}\left(\mathbf{B}^H\mathbf{B}\right) \\
&= \frac{1}{2}\operatorname{Tr}\left(\mathbf{A}^H\mathbf{B}\right) + \frac{1}{2}\operatorname{Tr}\left(\mathbf{B}^H\mathbf{A}\right) \overset{(a)}{=} \operatorname{Tr}(\mathbf{AB}),
\end{aligned}$$

where (a) is due to the fact that $\mathbf{A}$ and $\mathbf{B}$ are Hermitian matrices. Equation (5.9b) can be proved in a similar manner, and we omit the details for brevity. □

Based on Lemma 5.1, the first and second non-convex terms in (5.8b) can be transformed into the following difference of convex (DC) functions:

$$\overline{\gamma}_k \mathrm{Tr}\left(\mathbf{Q}_k \mathbf{H}_k \mathbf{W}_{\overline{k}} \mathbf{H}_k^H\right) = \frac{\overline{\gamma}_k}{2} \left\| \mathbf{Q}_k + \mathbf{H}_k \mathbf{W}_{\overline{k}} \mathbf{H}_k^H \right\|_F^2 - \frac{\overline{\gamma}_k}{2} \left\| \mathbf{Q}_k \right\|_F^2 - \frac{\overline{\gamma}_k}{2} \left\| \mathbf{H}_k \mathbf{W}_{\overline{k}} \mathbf{H}_k^H \right\|_F^2 \triangleq \Upsilon_{k\overline{k}}, \tag{5.10a}$$

$$-\mathrm{Tr}\left(\mathbf{Q}_k \mathbf{H}_k \mathbf{W}_k \mathbf{H}_k^H\right) = \frac{1}{2} \left\| \mathbf{Q}_k - \mathbf{H}_k \mathbf{W}_k \mathbf{H}_k^H \right\|_F^2 - \frac{1}{2} \left\| \mathbf{Q}_k \right\|_F^2 - \frac{1}{2} \left\| \mathbf{H}_k \mathbf{W}_k \mathbf{H}_k^H \right\|_F^2 \triangleq \Pi_{kk}. \tag{5.10b}$$

Furthermore, the non-convex rank-one constraint (5.8d) can be equivalently written as the following equality constraint:

$$\left\| \mathbf{Q}_k \right\|_* - \left\| \mathbf{Q}_k \right\|_2 = 0, \forall k \in \{t, r\}, \tag{5.11}$$

where $\left\| \mathbf{Q}_k \right\|_* = \sum_i \sigma_i \left(\mathbf{Q}_k\right)$ and $\left\| \mathbf{Q}_k \right\|_2 = \sigma_1 \left(\mathbf{Q}_k\right)$ denote the nuclear norm and the spectral norm, respectively, and $\sigma_i \left(\mathbf{Q}_k\right)$ is the i-th largest singular value of matrix $\mathbf{Q}_k$. Note that for any $\mathbf{Q}_k \in \mathbb{H}^M$ and $\mathbf{Q}_k \succeq 0$, we always have $\left\| \mathbf{Q}_k \right\|_* - \left\| \mathbf{Q}_k \right\|_2 \geq 0$, where equality holds if and only if $\mathbf{Q}_k$ is a rank-one matrix. Therefore, equality constraint (5.11) is only met for rank-one matrices $\{\mathbf{Q}_k\}$.

Next, we employ the penalty method (Nocedal & Wright, 2006) to solve problem (5.8). By exploiting (5.10a), (5.10b), and (5.11), we obtain the following optimization problem:

$$\min_{\mathbf{W}_k, \mathbf{Q}_k, \boldsymbol{\beta}_k} \mathrm{Tr}\left(\mathbf{W}_t\right) + \mathrm{Tr}\left(\mathbf{W}_r\right) + \eta \sum_{k \in \{t, r\}} \left(\left\| \mathbf{Q}_k \right\|_* - \left\| \mathbf{Q}_k \right\|_2 \right) \tag{5.12a}$$

$$\text{s.t. } \Upsilon_{k\overline{k}} + \Pi_{kk} + \overline{\gamma}_k \sigma_k^2 \leq 0, \forall k \in \{t, r\}, \tag{5.12b}$$

$$(5.8c), (5.8e), (5.8f), (5.8g), \tag{5.12c}$$

where equality constraint (5.11) is relaxed to a penalty term added to the objective function, and $\eta > 0$ is the penalty factor that penalizes the objective function if $\{\mathbf{Q}_k\}$ is not rank one. It can be verified that, when $\eta \rightarrow +\infty$, the solution $\{\mathbf{Q}_k\}$ of problem (5.12) always satisfies equality constraint (5.11), i.e., problems (5.8) and (5.12) are equivalent (Ben-Tal & Zibulevsky, 1997). However, if the initial value of the penalty factor η is chosen too large, the objective function of (5.12) is dominated by the penalty term, and the impact of the desired power minimization on the solution becomes negligible. To avoid this, we first initialize η with a small value to find a good starting point, and then, in the course of several iterations, we gradually increase η to a sufficiently large value to eventually obtain feasible rank-one matrices. Note that, for any given penalty factor $\eta > 0$, problem (5.12) is still non-convex due to the non-convexity of the objective function and the non-convex constraints (5.12b) and (5.8e). In the following, we employ successive convex approximation (SCA) (Dinh & Diehl, 2010) to obtain a suboptimal solution of (5.8) in an iterative manner.

Note that the penalty term, $\Upsilon_{k\bar{k}}$, and Π_{kk} are in the form of DC functions. For a given point $\mathbf{Q}_k^{(l)}$ in the l-th iteration of the SCA method, using first-order Taylor expansion, a convex upper bound for the penalty term can be obtained as follows:

$$\|\mathbf{Q}_k\|_* - \|\mathbf{Q}_k\|_2 \le \|\mathbf{Q}_k\|_* - \overline{\mathbf{Q}}_k^{(l)}, \tag{5.13}$$

where $\overline{\mathbf{Q}}_k^{(l)} \triangleq \left\|\mathbf{Q}_k^{(l)}\right\|_2 + \mathrm{Tr}\left[\bar{\mathbf{u}}\left(\mathbf{Q}_k^{(l)}\right)\left(\bar{\mathbf{u}}\left(\mathbf{Q}_k^{(l)}\right)\right)^H\left(\mathbf{Q}_k - \mathbf{Q}_k^{(l)}\right)\right]$ and $\bar{\mathbf{u}}\left(\mathbf{Q}_k^{(l)}\right)$ denotes the eigenvector corresponding to the largest eigenvalue of $\mathbf{Q}_k^{(l)}$.

Similarly, for given points $\left\{\mathbf{Q}_k^{(l)}, \mathbf{W}_k^{(l)}\right\}$ in the l-th iteration of the SCA method, convex upper bounds of $\Upsilon_{k\bar{k}}$ and Π_{kk} are respectively given by

$$\begin{aligned}\Upsilon_{k\bar{k}} \le {} & \frac{\overline{\gamma}_k}{2}\left\|\mathbf{Q}_k + \mathbf{H}_k\mathbf{W}_{\bar{k}}\mathbf{H}_k^H\right\|_F^2 + \frac{\overline{\gamma}_k}{2}\left\|\mathbf{Q}_k^{(l)}\right\|_F^2 - \overline{\gamma}_k\mathrm{Tr}\left(\left(\mathbf{Q}_k^{(l)}\right)^H\mathbf{Q}_k\right)\\ & + \frac{\overline{\gamma}_k}{2}\left\|\mathbf{H}_k\mathbf{W}_{\bar{k}}^{(l)}\mathbf{H}_k^H\right\|_F^2 - \overline{\gamma}_k\mathrm{Tr}\left(\left(\mathbf{H}_k^H\mathbf{H}_k\mathbf{W}_{\bar{k}}^{(l)}\mathbf{H}_k^H\mathbf{H}_k\right)^H\mathbf{W}_{\bar{k}}\right) \triangleq \left[\Upsilon_{k\bar{k}}\right]^{ub},\end{aligned} \tag{5.14a}$$

$$\begin{aligned}\Pi_{kk} \le {} & \frac{1}{2}\left\|\mathbf{Q}_k - \mathbf{H}_k\mathbf{W}_k\mathbf{H}_k^H\right\|_F^2 + \frac{1}{2}\left\|\mathbf{Q}_k^{(l)}\right\|_F^2 - \mathrm{Tr}\left(\left(\mathbf{Q}_k^{(l)}\right)^H\mathbf{Q}_k\right)\\ & + \frac{1}{2}\left\|\mathbf{H}_k\mathbf{W}_k^{(l)}\mathbf{H}_k^H\right\|_F^2 - \mathrm{Tr}\left(\left(\mathbf{H}_k^H\mathbf{H}_k\mathbf{W}_k^{(l)}\mathbf{H}_k^H\mathbf{H}_k\right)^H\mathbf{W}_k\right) \triangleq [\Pi_{kk}]^{ub}.\end{aligned} \tag{5.14b}$$

As a result, for given points $\left\{\mathbf{Q}_k^{(l)}, \mathbf{W}_k^{(l)}\right\}$, by replacing the non-convex terms with their convex upper bounds, problem (5.12) is transformed into the following optimization problem:

$$\min_{\mathbf{W}_k, \mathbf{Q}_k, \beta_k} \mathrm{Tr}\left(\mathbf{W}_t\right) + \mathrm{Tr}\left(\mathbf{W}_r\right) + \eta\sum\nolimits_{k\in\{t,r\}}\left(\|\mathbf{Q}_k\|_* - \overline{\mathbf{Q}}_k^{(l)}\right) \tag{5.15a}$$

$$\text{s.t. } \left[\Upsilon_{k\bar{k}}\right]^{ub} + [\Pi_{kk}]^{ub} + \overline{\gamma}_k\sigma_k^2 \le 0, \forall k \in \{t, r\}, \tag{5.15b}$$

$$\text{(5.8c), (5.8e), (5.8f), (5.8g).} \tag{5.15c}$$

Now, the remaining non-convexity of (5.15) is the non-convex rank-one constraint (5.8e). To handle this issue, we employ semidefinite relaxation (SDR) and solve the relaxed problem by ignoring (5.8e). The tightness of the relaxed version of problem (5.15) is shown in the following theorem.

Theorem 5.1 *Without loss of optimality, the solutions* $\{\mathbf{W}_k\}$ *obtained for the relaxed version of problem of* (5.15), *i.e., after dropping rank-one constraint* (5.8e), *always satisfy* Rank $(\mathbf{W}_k) = 1, \forall k \in \{t, r\}$.

Proof See Appendix C. □

The relaxed problem (5.15), which is a standard convex semidefinite program (SDP), can be efficiently solved via standard convex problem solvers such as CVX (Grant & Boyd, 2014). Then, we propose a penalty-based iterative algorithm for solving problem (5.8), which comprises two loops. In the outer loop, the penalty factor is gradually increased from one iteration to the next as follows: $\eta = \omega\eta$, where $\omega > 1$. The algorithm terminates when the penalty term satisfies the following criterion:

$$\max\left\{\|\mathbf{Q}_k\|_* - \|\mathbf{Q}_k\|_2, \forall k \in \{t, r\}\right\} \leq \varepsilon_1, \tag{5.16}$$

where ε_1 denotes a predefined maximum violation of equality constraint (5.11). Therefore, (5.11) will be eventually satisfied with accuracy ε_1 as η increases. In the inner loop, $\{\mathbf{Q}_k, \mathbf{W}_k\}$ are jointly optimized by iteratively solving the relaxed version of problem (5.15) for the given penalty factor. The objective function value of the relaxed version of (5.15) is nonincreasing in each iteration of the inner loop, and the optimal objective function value of the relaxed version of (5.15) is bounded from below. Therefore, as η approaches infinity, the developed penalty-based iterative algorithm is guaranteed to converge to a stationary point of the original problem (5.8) (Dinh & Diehl, 2010). The details of the developed algorithm are summarized in Algorithm 1.

The computational complexity of Algorithm 1 is analyzed as follows. The main complexity is caused by solving the relaxed version of problem (5.15) in the inner loop. As the relaxed problem is a standard SDP, the computational complexity for solving this problem is $\mathcal{O}\left(KM^{3.5} + 2N^{3.5}\right)$ if the interior-point method is employed (Luo et al., 2010), where $K = 2$ is the number of users. Then, the overall computational complexity of Algorithm 1 is $\mathcal{O}\left(I_{\text{out}} I_{\text{inn}}\left(KM^{3.5} + 2N^{3.5}\right)\right)$, where I_{inn} and I_{out} denote the number of inner and outer iterations required for convergence, respectively.

5.2.3 Numerical Results

In this section, numerical results are provided to validate the effectiveness of the proposed STARS-aided communication system.

Algorithm 1 Proposed penalty-based iterative algorithm for solving problem (5.8)

```
Initialize feasible points {Q_k^(0), W_k^(0)}, the penalty factor η.
repeat: outer loop
    Set iteration index l = 0 for inner loop.
    repeat: inner loop
        For given {Q_k^(l), W_k^(l)}, solve the relaxed version of problem (5.15).
        Update {Q_k^(l+1), W_k^(l+1)} with the obtained optimal solutions, and l = l + 1.
    until the fractional decrease of the objective function value is below a predefined threshold
    ϵ_1 > 0 or the maximum number of inner iterations l_max is reached.
    Update {Q_k^(0), W_k^(0)} with the current solutions {Q_k^(l), W_k^(l)}.
    Update η = ωη.
until the constraint violation is below a predefined threshold ε_1 > 0.
```

Fig. 5.3 The simulated setup

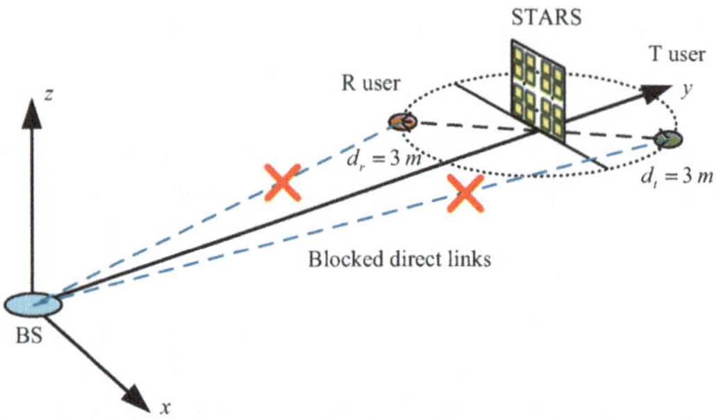

5.2.3.1 Simulation Setup

Figure 5.3 illustrates the considered 3D simulation setup, where the BS and the STARS are located at (0, 0, 0) meters and (0, 50, 0) meters, respectively. The T and R users are randomly located on half-circles centered at the STARS with a radius of $d_t = d_r = 3$ m. The STARS is assumed to be equipped with a uniform planar array (UPA) composed of $N = N_h N_v$ elements, where $N_h = 5$ denotes the number of elements along the horizontal plane and we increase N_v linearly with N. For our simulations, the narrowband quasi-static fading channels from the BS to the STARS and from the STARS to the two users are modeled as Rician fading channels as follows:

$$\mathbf{G} = \sqrt{\frac{\rho_0}{d_{BR}^{\alpha_{BR}}}} \left(\sqrt{\frac{K_{BR}}{K_{BR}+1}} \mathbf{G}^{\mathrm{LoS}} + \sqrt{\frac{1}{K_{BR}+1}} \mathbf{G}^{\mathrm{NLoS}} \right), \tag{5.17a}$$

$$\mathbf{v}_k = \sqrt{\frac{\rho_0}{d_{RU,k}^{\alpha_{RU}}}} \left(\sqrt{\frac{K_{RU}}{K_{RU}+1}} \mathbf{v}_k^{\mathrm{LoS}} + \sqrt{\frac{1}{K_{RU}+1}} \mathbf{v}_k^{\mathrm{NLoS}} \right), \tag{5.17b}$$

Table 5.2 System parameters

Parameter	Description	Value
α_{BR}	Path loss exponent for BS-STARS channels	2.2
α_{RU}	Path loss exponent for STARS-user channels	2.2
K_{BR}	Rician factor for BS-STARS channels	3 dB
K_{BU}	Rician factor for STARS-user channels	3 dB
ρ_0	Path loss at 1 meter	−30 dB
σ_k^2	User noise power	−90 dBm
η, χ	Initialized penalty factors for Algorithm 1	10^{-4}
ω, ϖ	Scaling factors for Algorithm 1	10
ϵ_1, ϵ_2	Convergence tolerance for SCA	10^{-2}
ε_1, ε_2	Accuracy for equality constraints	10^{-7}
$l_{\max}$	Maximum number of inner iterations of Algorithm 1	30

where d_{BR} and $d_{RU,k}$ denote the distances between the BS and the STARS and between the STARS and user $k \in \{t, r\}$, α_{BR} and α_{RU} denote the corresponding path loss exponents, K_{BR} and K_{RU} denote the Rician factors, ρ_0 represents the path loss at a reference distance of 1 meter, $\mathbf{G}^{\mathrm{LoS}}$ and $\mathbf{v}_k^{\mathrm{LoS}}$ are the deterministic LoS components, and $\mathbf{G}^{\mathrm{NLoS}}$ and $\mathbf{v}_k^{\mathrm{NLoS}}$ are the random NLoS components modeled as Rayleigh fading. The adopted system parameters are presented in Table 5.2. Without loss of generality, we assume that the two users for unicast transmission have the same QoS requirements, i.e., $\overline{R}_t = \overline{R}_r = \overline{R}_0 \triangleq \log_2\left(1+\overline{\gamma}_0\right)$ and $\overline{\gamma}_0$ is the minimum required signal-to-interference-plus-noise ratio (SINR). The communication requirement for multicast transmission is set to $\overline{R}_c \triangleq \log_2\left(1+\overline{\gamma}_c\right)$, where $\overline{\gamma}_c$ is the minimum required SINR. The power consumption results shown (i.e., Figs. 5.4 and 5.5) were obtained by averaging over 100 channel realizations. Specifically, we first randomly generated 100 user distributions and then obtained the corresponding channel realizations.

5.2.3.2 Baseline Schemes

To verify the effectiveness of the proposed STARS concept and the corresponding operating protocols, we compare the following two baseline schemes:

- **Baseline scheme 1 (also referred to as conventional RISs)**: In this case, the full-space coverage facilitated by the STARS in Fig. 5.3 is achieved by employing one conventional reflecting-only RIS and one transmitting-only RIS. The two conventional RISs are deployed adjacent to each other at the same location as the STARS. For a fair comparison, each conventional reflecting/transmitting RIS is assumed to have $N/2$ elements, where N is assumed to be an even number for simplicity. This baseline scheme can be regarded as a special case of an MS

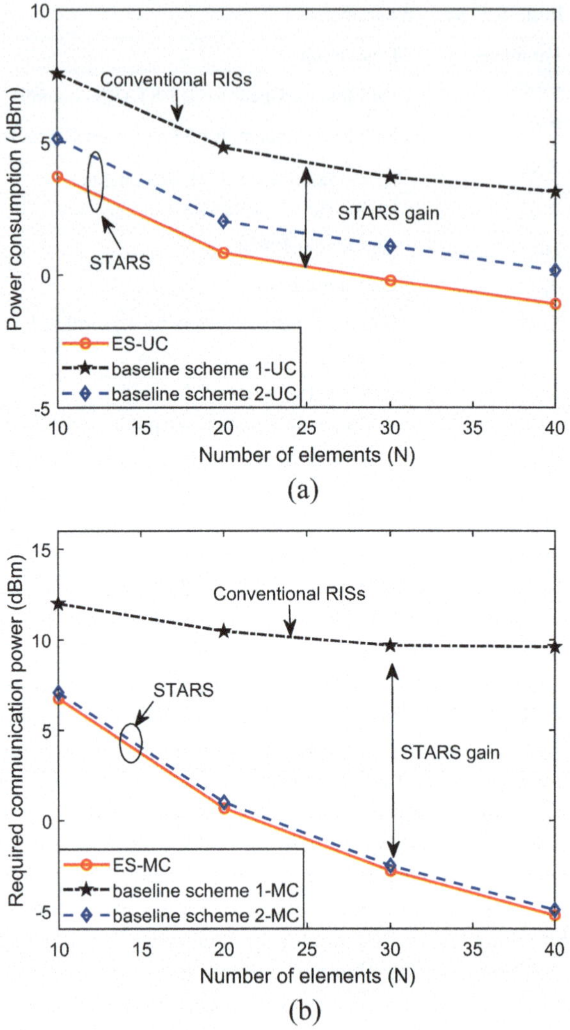

Fig. 5.4 Power consumption versus N for $M = 2$. (**a**) Unicast communication, $\overline{\gamma}_0 = 0$ dB. (**b**) Multicast communication, $\overline{\gamma}_c = 10$ dB

STARS, where $N/2$ elements operate in the T mode and $N/2$ elements operate in the R mode. Therefore, the resulting optimization problem can be solved by applying **Algorithm 1** with $\boldsymbol{\beta}^t = [\mathbf{1}_{1\times N/2}\ \mathbf{0}_{1\times N/2}]$ and $\boldsymbol{\beta}^r = [\mathbf{0}_{1\times N/2}\ \mathbf{1}_{1\times N/2}]$.

- **Baseline scheme 2 (also referred to as uniform energy splitting (UES))**: In this case, we assume that all elements of the ES STARS employ the same amplitude coefficients for transmission and reflection, respectively, i.e., $\beta_n^k = \overline{\beta}^k, \beta_n^r = \overline{\beta}^r \forall n \in \mathcal{N}$, where $0 \leq \overline{\beta}^k, \overline{\beta}^r \leq 1$ and $\overline{\beta}^t + \overline{\beta}^r = 1$. UES can be regarded as a special case of the ES STARS employing a group/surface-wise amplitude design. The resulting optimization problem can be solved by applying **Algorithm 1** with the above linear equality constraints.

Fig. 5.5 Power consumption versus M for $N = 10$. (**a**) Unicast communication, $\overline{\gamma}_0 = 0\,\text{dB}$. (**b**) Multicast communication, $\overline{\gamma}_c = 10\,\text{dB}$

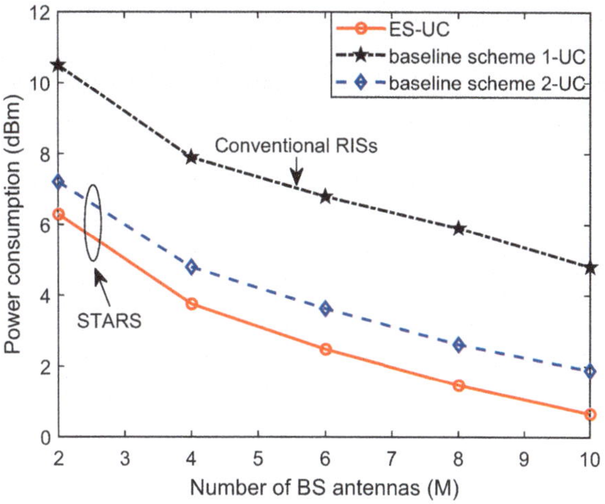

(a)

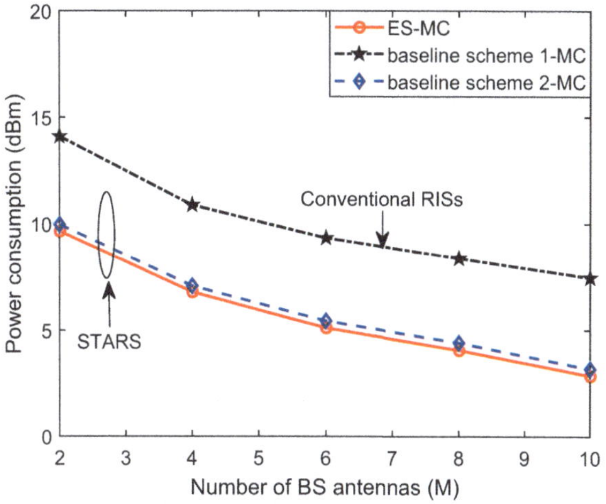

(b)

5.2.3.3 Required Power Consumption versus the Number of RIS Elements

In Fig. 5.4, we investigate the required power consumption versus the number of RIS elements, N, for both unicast and multicast transmission. Besides the three proposed operating protocols, the results obtained for the two baseline schemes are provided for comparison. We set $M = 2$, $\overline{\gamma}_0 = 0$ dB for unicast communication, and $M = 2$, $\overline{\gamma}_c = 10$ dB for multicast communication. As can be seen from Fig. 5.4, the required power consumption for all schemes and scenarios decreases as N increases. This is expected since larger N enables a higher transmission/reflection beamforming gain, which in turn reduces the required power consumption of the BS. Regarding the performance comparison between STARS and baseline scheme 1, as can be observed from Fig. 5.4, independent of the adopted operating protocol, STARS always outperforms conventional RISs in both unicast and multicast scenarios. The reasons behind this can be explained as follows. Since conventional RISs employ fixed numbers of transmission and reflection elements, they cannot exploit the same DoFs as available to the STARS to enhance the desired signal strength and/or mitigate inter-user interference. Therefore, conventional RISs always suffer from the worst performance. This limitation also causes the performance gap between conventional RISs and STARS to become more pronounced as N increases. The above performance comparison confirms the effectiveness of employing STARS in wireless communication systems.

For baseline scheme 2, a special case of ES, it is interesting to observe that there is a noticeable performance gap between ES and UES for unicast communication in Fig. 5.4a. However, the performance gap is negligible for multicast communication in Fig. 5.4b. Recall that the difference between ES and UES only lies in the element-wise and group-wise transmission and reflection amplitude control. The extra DoFs provided by the element-wise amplitude control allow ES to achieve an improved desired signal enhancement and inter-user interference mitigation compared to UES for unicast communication, thus achieving higher performance. For multicast communication, since there is no inter-user interference to be mitigated, the performance gain caused by the element-wise amplitude control vanishes. On the one hand, this result underscores the importance of element-wise transmission and reflection amplitude control for mitigating inter-user interference for unicast communication. On the other hand, the result also implies that UES is a promising operating protocol for multicast communication, where it achieves similar performance as ES. However, compared with ES, the group-wise amplitude control of UES reduces the overhead caused by exchanging configuration information between the BS and the STARS, which is a major bottleneck for the practical implementation of RISs.

5.2.3.4 Required Power Consumption versus the Number of BS Antennas

In Fig. 5.5, we study the required power consumption versus the number of BS antennas, N, by STARS and the baselines for both unicast and multicast communication. We set $M = 10$, $\overline{\gamma}_0 = 0$ dB for unicast communication, and $M = 10$, $\overline{\gamma}_c = 10$ dB for multicast communication. Figure 5.5 shows that the required power consumption for all schemes decreases as N increases thanks to a higher active beamforming gain. For unicast communication in Fig. 5.5a, the proposed STARS outperforms conventional RISs since more DoFs for transmission and reflection design can be exploited. As we consider a small N, for multicast communication in Fig. 5.5b, the loss in DoFs for employing conventional RISs is not significant. As can be observed in Fig. 5.5a, there is a considerable performance gap between UES and ES for unicast communication, while UES and ES achieve a similar performance for multicast communication in Fig. 5.5b. This is consistent with the results in Fig. 5.4.

5.3 STARS Beamforming Design Under the Coupled Phase-Shift Model

In this section, we continue to study the STARS beamforming design with the coupled phase-shift constraints, where an iterative element-wise alternating optimization algorithm is proposed.

5.3.1 *System Model and Problem Formulation*

5.3.1.1 System Model

We consider a narrowband STARS-aided downlink communication system operating over frequency-flat channels, where a single-antenna access point (AP) transmits to two single-antenna users with the aid of a STARS comprising N elements, as shown in Fig. 5.6. In this section, we assume that one user is located in the transmission space of the STARS, referred to as the T user, and the other is located in the reflection space of the STARS, referred to as the R user. Let $d_k \in \mathbb{C}^{1\times 1}$ and $\mathbf{v}_k \in \mathbb{C}^{N\times 1}$ denote the channels from the AP to user k and the STARS to user k, respectively. Here, $k \in \{t, r\}$ indicates the T and R users. Let $\mathbf{g} \in \mathbb{C}^{N\times 1}$ denote the channel between the AP and the STARS. To characterize the maximum performance gain of STARS, the perfect CSI of all channels is assumed to be available at the AP.

For the studied communication system, we consider both NOMA and OMA. In NOMA, users share the same time/frequency resources and employ SIC to detect the superimposed signal delivered by the AP. In OMA, the users detect their intended signals independently by exploiting orthogonal time/frequency resources of equal size.

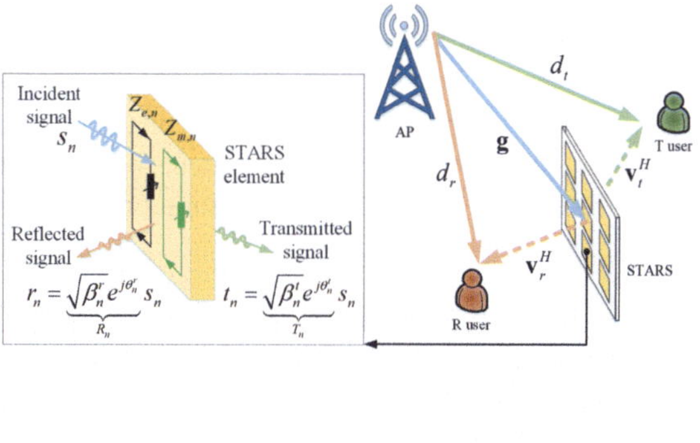

Fig. 5.6 Illustration of a STARS-aided downlink communication system

5.3.1.2 NOMA

Let x_k, p_k, and $c_k \triangleq d_k + \mathbf{v}_k^H \mathbf{\Theta}_k \mathbf{g}$ denote the information-bearing symbol, the allocated power, and the effective channel for user $k \in \{t, r\}$, respectively. The superimposed signal received at user $k \in \{t, r\}$ is given by

$$y_k = c_k \left(p_t x_t + p_r x_r\right) + n_k, \tag{5.18}$$

where $n_k \sim \mathcal{CN}\left(0, \sigma^2\right)$ represents the AWGN. In downlink NOMA, the user having the higher channel power gain (referred to as the strong user) first detects the signal of the user having the lower channel power gain (referred to as the weak user). Then, the strong user will subtract the detected weak user's signal from the received signal before detecting its own signal. Therefore, the achievable communication rate of user $k \in \{t, r\}$ is given by

$$R_k^{\mathrm{N}} = \log_2 \left(1 + \frac{|c_k|^2 p_k}{\lambda_k |c_k|^2 p_{\overline{k}} + \sigma^2}\right), \tag{5.19}$$

where $\overline{k} = r$, if $k = t$; and $\overline{k} = t$, otherwise. Here, binary variables, $\lambda_k \in \{0, 1\}, k \in \{t, r\}$, indicate the decoding order of the two users and $\lambda_t + \lambda_s = 1$. If $|c_t|^2 \geq |c_r|^2$ (i.e., T user is the strong user), we have $\lambda_t = 0$ and $\lambda_r = 1$; otherwise, $\lambda_t = 1$ and $\lambda_r = 0$.

5.3.1.3 OMA

By employing orthogonal time/frequency resources, OMA avoids inter-user interference. Therefore, the corresponding achievable communication rate of user $k \in \{t, r\}$ is given by

$$R_k^{\mathrm{O}} = \frac{1}{2} \log_2 \left(1 + \frac{|c_k|^2 p_k}{\frac{1}{2}\sigma^2}\right). \tag{5.20}$$

5.3.1.4 Problem Formulation

Our goal is to minimize the total power consumption of the AP by jointly optimizing the TRCs, subject to the rate requirements of both users and the coupled constraints on the TRCs. Therefore, the optimization problem can be formulated as follows:

$$\min_{p_k, \mathbf{\Theta}_k, \lambda_k} \quad p_t + p_r \tag{5.21a}$$

$$\text{s.t. } R_k^{\mathrm{X}} \geq \overline{R}_k, \forall k \in \{t, r\}, \tag{5.21b}$$

$$\beta_n^t + \beta_n^r = 1, \forall n \in \mathcal{N}, \tag{5.21c}$$

$$\left|\theta_n^t - \theta_n^r\right| = \frac{1}{2}\pi \text{ or } \frac{3}{2}\pi, \forall n \in \mathcal{N}, \tag{5.21d}$$

$$\theta_n^t, \theta_n^r \in [0, 2\pi), \beta_n^t, \beta_n^r \in [0, 1], \forall n \in \mathcal{N}, \tag{5.21e}$$

$$\begin{cases} \lambda_t = 0, \lambda_r = 1, \text{ if } |c_t|^2 \geq |c_r|^2, \\ \lambda_t = 1, \lambda_r = 0, \text{ otherwise}, \end{cases} \tag{5.21f}$$

$$\lambda_t, \lambda_r \in \{0, 1\}, \lambda_t + \lambda_r = 1, \tag{5.21g}$$

where $\overline{R}_k$ denotes the minimum rate requirement of user k and $\mathrm{X} \in \{\mathrm{N}, \mathrm{O}\}$ indicates whether NOMA or OMA is employed. Constraints (5.21c) and (5.21d) are the coupled amplitude and phase-shift coefficients, respectively. Variable λ_k and constraints (5.21f) and (5.21g) are only valid when NOMA is employed.

5.3.2 *Iterative Element-Wise Alternating Optimization Algorithm*

To solve problem (5.21), we first transform it into a tractable form. Note that for the optimal solution of (5.21), the inequality rate constraint (5.21b) has to be satisfied with equality. Therefore, problem (5.21) can be rewritten as follows:

$$\min_{\boldsymbol{\Theta}_k, \lambda_k} \quad P^{\mathrm{X}} \tag{5.22a}$$

$$\text{s.t. } (5.21\text{c})\text{–}(5.21\text{g}), \tag{5.22b}$$

where P^{X} for NOMA and OMA is respectively given by

$$P^{\mathrm{N}} = \begin{cases} \dfrac{\overline{\gamma}_r^{\mathrm{N}}}{|c_r|^2} + \dfrac{\overline{\gamma}_t^{\mathrm{N}}\left(\overline{\gamma}_r^{\mathrm{N}} + 1\right)}{|c_t|^2} \triangleq P^{\mathrm{N}}_{\lambda_t=0,\lambda_r=1}, \\ \dfrac{\overline{\gamma}_t^{\mathrm{N}}}{|c_t|^2} + \dfrac{\overline{\gamma}_r^{\mathrm{N}}\left(\overline{\gamma}_t^{\mathrm{N}} + 1\right)}{|c_r|^2} \triangleq P^{\mathrm{N}}_{\lambda_t=1,\lambda_r=0}, \end{cases} \tag{5.23a}$$

$$P^{\mathrm{O}} = \frac{\overline{\gamma}_t^{\mathrm{O}}}{|c_t|^2} + \frac{\overline{\gamma}_r^{\mathrm{O}}}{|c_r|^2}, \tag{5.23b}$$

where $\overline{\gamma}_k^{\mathrm{N}} \triangleq \left(2^{\overline{R}_k} - 1\right)\sigma^2$ and $\overline{\gamma}_t^{\mathrm{O}} \triangleq \left(2^{2\overline{R}_k} - 1\right)\sigma^2/2$, $\forall k \in \{t, r\}$. As can be observed, objective functions for NOMA and OMA share a similar structure, while NOMA requires additional decoding order constraints (5.21f) and (5.21g). As a result, in the following, we mainly focus on solving problem (5.22) for NOMA since the proposed algorithm is also applicable to OMA. Recall that there are $2! = 2$ possible decoding orders for problem (5.22) for NOMA. The optimal solution can be obtained by exhaustively searching over the two options, i.e., $P^{\mathrm{N}} = \min\left(P^{\mathrm{N}}_{\lambda_t=0,\lambda_r=1}, P^{\mathrm{N}}_{\lambda_t=1,\lambda_r=0}\right)$. Before solving (5.22), we first introduce the following proposition.

Proposition 5.1 *For the optimal solution of problem* (5.22)*, the NOMA decoding order constraints* (5.21f) *and* (5.21g) *are automatically satisfied.*

Proof Let $\left\{\boldsymbol{\Theta}_k^{op}, k \in \{t, r\}\right\}$ denote the optimal solution of problem (5.22). Then, we have

$$\begin{aligned} & P^{\mathrm{N}}_{\lambda_t=0,\lambda_r=1}\left(\{\boldsymbol{\Theta}_k^{op}\}\right) - P^{\mathrm{N}}_{\lambda_t=1,\lambda_r=0}\left(\{\boldsymbol{\Theta}_k^{op}\}\right) \\ = & \overline{\gamma}_t\overline{\gamma}_r\left(\frac{1}{\left|c_t\left(\{\boldsymbol{\Theta}_k^{op}\}\right)\right|^2} - \frac{1}{\left|c_r\left(\{\boldsymbol{\Theta}_k^{op}\}\right)\right|^2}\right) \triangleq \delta, \end{aligned} \tag{5.24}$$

where $\left|c_t\left(\{\boldsymbol{\Theta}_k^{op}\}\right)\right|^2$ and $\left|c_r\left(\{\boldsymbol{\Theta}_k^{op}\}\right)\right|^2$ denote the corresponding effective channel power gain for the T and R users, respectively. If $\lambda_t = 0, \lambda_r = 1$ is the optimal decoding order, $\left|c_t\left(\{\boldsymbol{\Theta}_k^{op}\}\right)\right|^2 \geq \left|c_r\left(\{\boldsymbol{\Theta}_k^{op}\}\right)\right|^2$ has to hold such that $\delta \leq 0$ (i.e., $P^{\mathrm{N}}_{\lambda_t=0,\lambda_r=1}\left(\{\boldsymbol{\Theta}_k^{op}\}\right) \leq P^{\mathrm{N}}_{\lambda_t=1,\lambda_r=0}\left(\{\boldsymbol{\Theta}_k^{op}\}\right)$). Otherwise, $\lambda_t = 0, \lambda_r = 1$ cannot be the optimal decoding order. The case of $\lambda_t = 1, \lambda_r = 0$ can be proved in a similar manner. □

By exploiting Proposition 5.1, we can remove constraints (5.21f) and (5.21g) from problem (5.22). In the following, for simplicity of presentation, we consider the case $\lambda_t = 0, \lambda_r = 1$ as an example. The resulting optimization problem simplifies to[1]

$$\min_{\boldsymbol{\Theta}_k} \ \frac{\overline{\gamma}_r^{\mathrm{N}}}{|c_r|^2} + \frac{\overline{\gamma}_t^{\mathrm{N}}\left(\overline{\gamma}_r^{\mathrm{N}}+1\right)}{|c_t|^2} \tag{5.25a}$$

$$\text{s.t. } (5.21\text{c}), (5.21\text{d}), (5.21\text{e}). \tag{5.25b}$$

However, it is still challenging to solve problem (5.25), since not only the objective function is non-convex with respect to $\boldsymbol{\Theta}_k$ but also the feasible set of the TRCs. Since it is difficult to find a globally optimal solution for such a challenging problem, in the following, we decompose (5.25) into two subproblems, namely a phase-shift coefficient design problem and an amplitude coefficient design problem. For each

[1] For NOMA with $\lambda_t = 1, \lambda_r = 0$ and OMA, we just need to replace the objective function of (5.25) with $P^{\mathrm{N}}_{\lambda_t=1,\lambda_r=0}$ and P^{O} from (5.23).

subproblem, an element-wise alternating optimization algorithm will be developed to find a high-quality suboptimal solution.

5.3.2.1 Phase-Shift Coefficient Design

To start with, we first define transmission/reflection amplitude and phase-shift vectors as $\boldsymbol{\beta}_k \triangleq \left[\sqrt{\beta_1^k}, \sqrt{\beta_2^k}, \ldots, \sqrt{\beta_N^k}\right]^T$ and $\mathbf{q}_k = \left[e^{j\theta_1^k}, e^{j\theta_2^k}, \ldots, e^{j\theta_N^k}\right]^T, \forall k \in \{t, r\}$, respectively. Therefore, the transmission/reflection-coefficient matrix can be reexpressed as $\boldsymbol{\Theta}_k = \operatorname{diag}(\boldsymbol{\beta}_k) \operatorname{diag}(\mathbf{q}_k)$. Accordingly, the effective channel power gain of user $k \in \{t, r\}$ can be rewritten as follows:

$$|c_k|^2 = \left|d_k + \mathbf{v}_k^H \operatorname{diag}(\boldsymbol{\beta}_k) \operatorname{diag}(\mathbf{q}_k)\,\mathbf{g}\right|^2 = \left|\mathbf{s}_k^H \overline{\mathbf{q}}_k\right|^2, \tag{5.26}$$

where $\mathbf{s}_k^H \triangleq \left[\mathbf{v}_k^H \operatorname{diag}(\boldsymbol{\beta}_k) \operatorname{diag}(\mathbf{g})\ d_k\right] \in \mathbb{C}^{1\times(N+1)}$ and $\overline{\mathbf{q}}_k \triangleq \left[\mathbf{q}_k^T\ 1\right]^T \in \mathbb{C}^{(N+1)\times 1}$. For given $\boldsymbol{\beta}_k$, problem (5.25) reduces to the following transmission and reflection phase-shift coefficient design problem:

$$\min_{\overline{\mathbf{q}}_k, \psi_k} \ \frac{\overline{\gamma}_r^{\mathrm{N}}}{\psi_r} + \frac{\overline{\gamma}_t^{\mathrm{N}}\left(\overline{\gamma}_r^{\mathrm{N}} + 1\right)}{\psi_t} \tag{5.27a}$$

$$\text{s.t.}\ \left|\mathbf{s}_k^H \overline{\mathbf{q}}_k\right|^2 \geq \psi_k, \forall k \in \{t, r\}, \tag{5.27b}$$

$$[\mathbf{q}_t]_n = j[\mathbf{q}_r]_n \text{ or } [\mathbf{q}_t]_n = -j[\mathbf{q}_r]_n, \forall n \in \mathcal{N}, \tag{5.27c}$$

$$\left|[\mathbf{q}_t]_n\right| = 1, \left|[\mathbf{q}_r]_n\right| = 1, \forall n \in \mathcal{N}, \tag{5.27d}$$

where $\{\psi_k\}$ are auxiliary variables, $j^2 = -1$, and $[\cdot]_n$ represents the n-th element of a vector. Here, constraint (5.27c) is equivalent to constraint (5.21d), where $[\mathbf{q}_t]_n = j[\mathbf{q}_r]_n$ implies that $\left|\theta_n^t - \theta_n^r\right| = \frac{1}{2}\pi$ and $[\mathbf{q}_t]_n = -j[\mathbf{q}_r]_n$ implies that $\left|\theta_n^t - \theta_n^r\right| = \frac{3}{2}\pi$. Although the objective function of (5.27) is convex, it is still a non-convex problem since the LHS of (5.27b) is not concave and constraints (5.27c) and (5.27d) are non-convex. To handle this obstacle, we propose an element-wise alternating optimization algorithm. For any one of the elements in $\mathbf{q}_k$ (i.e., $[\mathbf{q}_k]_n$) with the other $N-1$ elements (i.e., $\left\{[\mathbf{q}_k]_l\right\}_{l\neq n}^N$) fixed, $\left|\mathbf{s}_k^H \overline{\mathbf{q}}_k\right|^2$ can be expressed as follows:

$$\left|\mathbf{s}_k^H \overline{\mathbf{q}}_k\right|^2 = A_{k,n} + 2\,\mathrm{Re}\left\{B_{k,n}[\mathbf{q}_k]_n\right\}, \tag{5.28}$$

where $A_{k,n} \triangleq \left|\sum_{l\neq n}^{N+1}\left[\mathbf{s}_k^H\right]_l [\mathbf{q}_k]_l\right|^2 + \left|\left[\mathbf{s}_k^H\right]_n\right|^2$, $B_{k,n} \triangleq \left[\mathbf{s}_k^H\right]_n \left\{\sum_{l\neq n}^{N+1}\left[\mathbf{s}_k^H\right]_l [\mathbf{q}_k]_l\right\}^*$, and $[\mathbf{q}_k]_{N+1} \triangleq 1, \forall k \in \{t, r\}, n \in \mathcal{N}$. In this case, $\left|\mathbf{s}_k^H \overline{\mathbf{q}}_k\right|^2$ is an affine function with respect to $[\mathbf{q}_k]_n$. For given $N-1$ transmission and reflection phase-shift coefficients, the design of the n-th transmission and reflection phase-shift coefficients can be formulated as follows:

$$\min_{[\mathbf{q}_r]_n, \psi_k} \frac{\overline{\gamma}_r^{\mathrm{N}}}{\psi_r} + \frac{\overline{\gamma}_t^{\mathrm{N}}\left(\overline{\gamma}_r^{\mathrm{N}}+1\right)}{\psi_t} \tag{5.29a}$$

$$\text{s.t. } A_{r,n} + 2\,\mathrm{Re}\left\{B_{r,n}[\mathbf{q}_r]_n\right\} \geq \psi_r, \tag{5.29b}$$

$$A_{t,n} + 2\,\mathrm{Re}\left\{B_{t,n}\left\{\pm j[\mathbf{q}_r]_n\right\}\right\} \geq \psi_t, \tag{5.29c}$$

$$\left|[\mathbf{q}_r]_n\right| = 1. \tag{5.29d}$$

As there are two possible phase-shift differences between $[\mathbf{q}_t]_n$ and $[\mathbf{q}_r]_n$, we can first solve the two subproblems and then select the optimal solution from the two results. For each subproblem, the non-convexity only lies in the unit-modulus constraint (5.29d). To address this issue, we can relax the non-convex unit-modulus constraint (5.29d) into a convex one, i.e., $\left|[\mathbf{q}_r]_n\right| \leq 1$. By exploiting the result in (Zhang and Zhang, 2021, Appendix B), it can be shown that the optimal solution to the relaxed problem will always satisfy $\left|[\mathbf{q}_r]_n\right| = 1$. The relaxed problem is a standard convex optimization problem, which can be efficiently solved via standard convex problem solvers such as CVX (Grant & Boyd, 2014). By alternatingly optimizing each phase-shift coefficient $[\mathbf{q}_r]_n$ with all the other $\left\{[\mathbf{q}_k]_l\right\}_{l\neq n}^{N}$ fixed, a suboptimal solution to (5.27) can be obtained.

5.3.2.2 Amplitude Coefficient Design

Similarly, for optimizing the transmission and reflection amplitude coefficients, we first rewrite the effective channel power gain of user $k \in \{t, r\}$ as follows:

$$|c_k|^2 = \left|d_k + \mathbf{v}_k^H \mathrm{diag}\left(\boldsymbol{\beta}_k\right) \mathrm{diag}\left(\mathbf{q}_k\right) \mathbf{g}\right|^2 = \left|\mathbf{b}_k^H \overline{\boldsymbol{\beta}}_k\right|^2, \tag{5.30}$$

where $\mathbf{b}_k^H \triangleq \left[\mathbf{v}_k^H \mathrm{diag}\left(\mathbf{q}_k\right) \mathrm{diag}\left(\mathbf{g}\right)\ d_k\right] \in \mathbb{C}^{1\times(N+1)}$ and $\overline{\boldsymbol{\beta}}_k \triangleq \left[\boldsymbol{\beta}_k^T\ 1\right]^T \in \mathbb{C}^{(N+1)\times 1}$. For each β_n^k and fixing the remaining $N-1$ coefficients, $\left|\mathbf{b}_k^H \overline{\boldsymbol{\beta}}_k\right|^2$ can be expressed as follows:

$$\left|\mathbf{b}_k^H \overline{\boldsymbol{\beta}}_k\right|^2 = C_{k,n} + D_{k,n}\beta_n^k + E_{k,n}\sqrt{\beta_n^k}, \tag{5.31}$$

where $E_{k,n} \triangleq 2\,\mathrm{Re}\left\{\left[\mathbf{b}_k^H\right]_n \left\{\sum_{l\neq n}^{N+1} \left[\mathbf{b}_k^H\right]_l \sqrt{\beta_l^k}\right\}^*\right\}$, $D_{k,n} \triangleq \left|\left[\mathbf{b}_k^H\right]_n\right|^2$, $C_{k,n} \triangleq \left|\sum_{l\neq n}^{N+1} \left[\mathbf{s}_k^H\right]_l \sqrt{\beta_l^k}\right|^2$, and $\beta_{N+1}^k \triangleq 1, \forall k \in \{t, r\}, n \in \mathcal{N}$. For given $N-1$ transmission and reflection amplitude coefficients, the design of the n-th transmission and reflection amplitude coefficients can be formulated as the following problem:

$$\min_{\beta_n^k, \psi_k} \frac{\overline{\gamma}_r^{\mathrm{N}}}{\psi_r} + \frac{\overline{\gamma}_t^{\mathrm{N}}\left(\overline{\gamma}_r^{\mathrm{N}}+1\right)}{\psi_t} \tag{5.32a}$$

$$\text{s.t. } C_{k,n} + D_{k,n}\beta_n^k + E_{k,n}\sqrt{\beta_n^k} \geq \psi_k, \forall k \in \{t, r\}, \tag{5.32b}$$

$$\beta_n^t, \beta_n^r \in [0,1], \beta_n^t + \beta_n^r = 1, \forall n \in \mathcal{N}. \tag{5.32c}$$

The potential non-convexity of (5.32) lies in the third term of the LHS of (5.32b), whose convexity depends on the sign of $E_{k,n}$. If $E_{k,n} \geq 0$, the LHS of (5.32b) is concave with respect to β_n^k. In this case, problem (5.32) is convex and can be efficiently solved using CVX (Grant & Boyd, 2014). However, if $E_{k,n} < 0$, the LHS of (5.32b) is not concave but convex with respect to β_n^k and problem (5.32) is non-convex. To solve this issue, for a given feasible point $\beta_n^{k(0)}$, by employing the first-order Taylor expansion, a lower bound of $E_{k,n}\sqrt{\beta_n^k}$ can be obtained as follows:

$$E_{k,n}\sqrt{\beta_n^k} \geq \frac{E_{k,n}\left(\beta_n^k + \beta_n^{k(0)}\right)}{2\sqrt{\beta_n^{k(0)}}} \triangleq \Gamma_{k,n}\left(\beta_n^k, \beta_n^{k(0)}\right). \tag{5.33}$$

Therefore, for a given feasible point $\beta_n^{k(0)}$ and $E_{k,n} < 0$, problem (5.32) can be approximated as the following convex optimization problem:

$$\min_{\beta_n^k, \psi_k} \frac{\overline{\gamma}_r^{\mathrm{N}}}{\psi_r} + \frac{\overline{\gamma}_t^{\mathrm{N}}\left(\overline{\gamma}_r^{\mathrm{N}}+1\right)}{\psi_t} \tag{5.34a}$$

$$\text{s.t. } C_{k,n} + D_{k,n}\beta_n^k + \Gamma_{k,n}\left(\beta_n^k, \beta_n^{k(0)}\right) \geq \psi_k, \forall k \in \{t, r\}, \tag{5.34b}$$

$$\beta_n^t, \beta_n^r \in [0,1], \beta_n^t + \beta_n^r = 1, \forall n \in \mathcal{N}, \tag{5.34c}$$

which can be solved using CVX (Grant & Boyd, 2014) to find a suboptimal solution to the original problem (5.32). Similarly, we can alternatingly optimize each $\left\{\beta_n^k\right\}$ with all the other $\left\{\beta_l^k\right\}_{l\neq n}^N$ fixed to obtain a suboptimal solution to the amplitude coefficient design subproblem.

5.3.2.3 Overall Algorithm

Based on the above two subproblems, an iterative element-wise alternating optimization algorithm is developed for solving problem (5.25). In each iteration, the two transmission/reflection phase-shift and amplitude coefficient design

Algorithm 2 Element-wise alternating optimization algorithm for solving (5.25)

1: Randomly initialize feasible $\{\mathbf{q}_k, k \in \{t, r\}\}$ and initialize $\left\{\beta_n^t = \beta_n^r = 0.5, \forall n \in \mathcal{N}\right\}$.
2: **repeat**
3: Optimize and update the phase-shift coefficients in an element-wise manner by solving (5.29).
4: Optimize and update the amplitude coefficients in an element-wise manner by solving (5.32) or (5.34).
5: **until** the fractional decrease of the objective function value is below a predefined threshold.

subproblems are alternatingly optimized. Within each subproblem, each phase-shift/amplitude coefficient is alternatingly optimized in an element-wise manner. The details of the developed algorithm are summarized in Algorithm 2. The computational complexity of Algorithm 2 is analyzed as follows. The main complexity is caused by solving each element-wise phase-shift and amplitude coefficient design problem, which has a complexity of $\mathcal{O}(1)$ if the interior-point method is employed. As there are two possible phase-shift differences for (5.29), the corresponding complexity for the N-element transmission and reflection phase-shift coefficient design problem is $\mathcal{O}(2N)$. The complexity for the N-element transmission and reflection amplitude coefficient design problem is $\mathcal{O}(N)$. Therefore, the overall computational complexity of Algorithm 2 is $\mathcal{O}(I(3N))$, where I denotes the number of iterations required for convergence. As a result, the complexity for solving problem (5.22) for NOMA and OMA is $\mathcal{O}(2I(3N))$ and $\mathcal{O}(I(3N))$, respectively, since for NOMA two possible decoding orders have to be considered. Overall, the complexity of the proposed algorithm increases only linearly with N, which is of vital importance since the number of STAR elements is usually large in practice.

5.3.3 *Numerical Results*

In this section, numerical results are provided to compare the performance of the proposed algorithm with two benchmark schemes. The considered 3D simulation setup is shown in Fig. 5.7, where the locations of the BS and the STARS are (0, 0, 0) meter and (0, 50, 0) meter, respectively. The T and R users are randomly located on half-circles centered at the STARS with a radius of 3 m. The BS-user channel is modeled as Rayleigh fading channel with a path loss exponent of 3.5, while the BS-STARS and the STARS-user channels are modeled as Rician fading channels with a path loss exponent of 2.2 and a Rician factor of 3 dB. The path loss at a reference distance of 1 meter is set to -30 dB, and the noise power is set to $\sigma^2 = -80$ dBm. The following results were obtained by averaging over 100 channel realizations, where 100 user distributions were first randomly generated, and then the corresponding channel realizations were generated accordingly.

For performance comparison, we consider the following two benchmark schemes. (1) **An upper bound**, where the phase-shift coefficients of the STARS

Fig. 5.7 The simulated setup

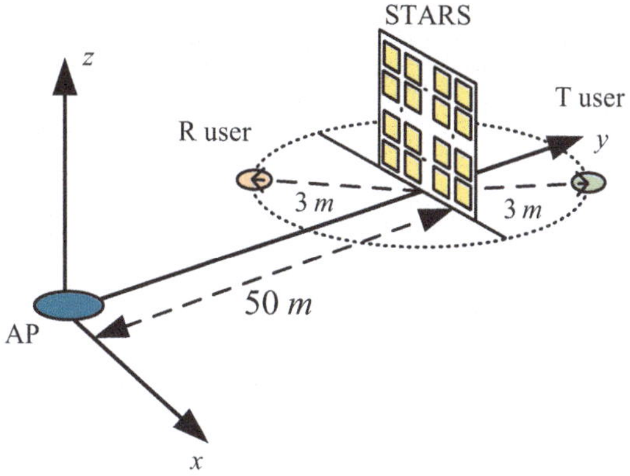

for transmission and reflection can be independently adjusted (Mu et al., 2022). (2) **Conventional RISs**, where one conventional $N/2$-element reflecting-only RIS and one $N/2$-element transmitting-only RIS are deployed adjacent to each other. For each benchmark scheme, both NOMA and OMA are considered.

In Fig. 5.8, the required power consumption versus the number of RIS elements is studied. We consider two cases, namely symmetric rate requirements ($\overline{R}_t = \overline{R}_r = 2$ bit/s/Hz) and asymmetric rate requirements ($\overline{R}_t = 5$ bit/s/Hz, $\overline{R}_r = 1$ bit/s/Hz). As can be observed from Fig. 5.5, for all cases and schemes, the required power consumption decreases as N increases due to the higher transmission/reflection gain. As each element for STARS has TRCs to be optimized (i.e., more DoFs to be exploited), STARS significantly outperform conventional RISs. Regarding the performance loss caused by the coupled phase-shift model, for both NOMA and OMA, a noticeable performance gap can be observed in Fig. 5.8a, while it becomes negligible in Fig. 5.8b. The reasons behind this are as follows. For symmetric rate requirements, STARS tend to maximize the effective channel power gains of both users to minimize power consumption. Therefore, each element has to be operated in the real STAR mode (i.e., $\beta_n^t \neq 0$ and $\beta_n^r \neq 0$), where the STARS employing the independent phase-shift model can exploit more DoFs than that employing the coupled phase-shift model. However, for asymmetric rate requirements, STARS tend to maximize the effective channel power gain of the user having the higher rate requirement (i.e., the T user) to greatly reduce the power consumption. As a result, each element is optimized to work in the pure transmission mode for serving the T user, while the R user is served only through the direct link from the BS. In this case, the effect of the coupled phase-shift model can be ignored. Moreover, it can be observed that the performance gain of NOMA over OMA is more pronounced in the case of asymmetric rate requirements compared to the case of symmetric rate requirements. This is expected since as explained above, STARS prefer to only enhance the channel power gain of the rate-hungry user for the case of asymmetric rate requirements, thus yielding a significant channel quality difference where NOMA achieves a higher performance gain.

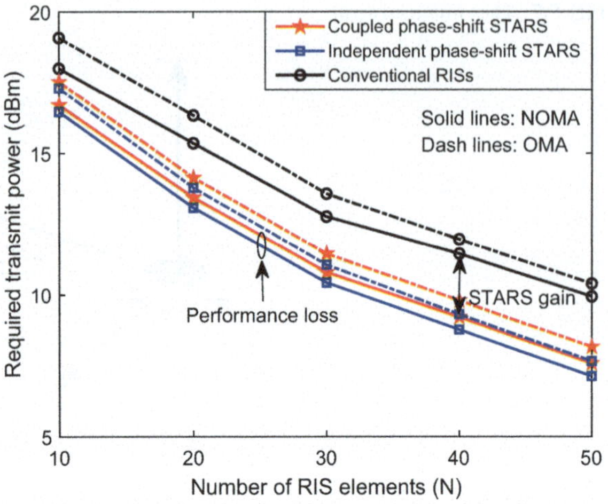

(a)

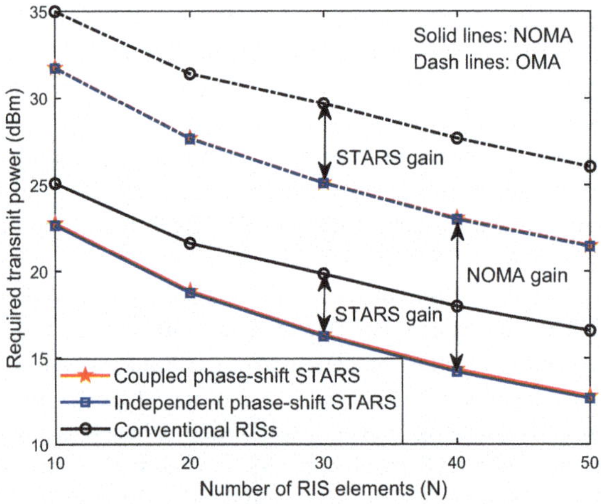

(b)

Fig. 5.8 Power consumption versus N. (**a**) $\overline{R}_t = \overline{R}_r = 2$ bit/s/Hz. (**b**) $\overline{R}_t = 5$ bit/s/Hz, $\overline{R}_r = 1$ bit/s/Hz

5.4 STARS Beamforming Design: A General Optimization Framework

To tackle the coupled phase-shift constraint, the element-wise alternating optimization algorithm proposed in Sect. 5.3 is only applicable to a system comprising a single-antenna AP and two single-antenna users. More importantly, for the aforementioned algorithm, the optimality of the obtained solution cannot be guaranteed. This motivates us to propose a general optimization framework with provable optimality for STARS beamforming design with the coupled phase-shift model.

5.4.1 A General Optimization Framework for STARS with Coupled Phase Shifts

Consider the following STARS beamforming optimization problem:

$$\min_{\mathbf{x}\in\mathcal{X},\boldsymbol{\theta}_t,\boldsymbol{\theta}_r} \quad F(\mathbf{x},\boldsymbol{\theta}_t,\boldsymbol{\theta}_r) \tag{5.35a}$$

$$\text{s.t.} \quad \beta_{t,n}^2+\beta_{r,n}^2=1, \forall n\in\mathcal{N}, \tag{5.35b}$$

$$\cos(\phi_{t,n}-\phi_{r,n})=0, \forall n\in\mathcal{N}, \tag{5.35c}$$

where $\mathcal{X}$ denotes the convex feasible set of optimization variable $\mathbf{x}$, vector $\boldsymbol{\theta}_i = [\beta_{i,1}e^{j\phi_{i,1}}, \ldots, \beta_{i,N}e^{j\phi_{i,N}}]^T, \forall i \in \{t,r\}$, contains the TRCs, and $F(\mathbf{x},\boldsymbol{\theta}_t,\boldsymbol{\theta}_r)$ is the convex objective function or the convex approximation of the original objective function for the specific problem considered. Depending on the application scenario, $F(\cdot)$ can be some utility function such as transmit power, weighted sum rate, harvested energy, and sensing accuracy, and $\mathbf{x}$ can represent the transmit waveforms or resource allocation variables. Note that in the above problem, the constraints $\beta_{t,n}\in[0,1]$ and $\beta_{r,n}\in[0,1]$ are omitted without changing the optimal objective value. This is because for any optimal solution $(\beta_{i,n}^\star, \phi_{i,n}^\star), \forall i\in\{t,r\}$, of problem (5.35) with $\beta_{i,n}^\star\in[-1,0]$, it can be verified that $(-\beta_{i,n}^\star, \phi_{i,n}^\star+\pi)$ is also a feasible solution achieving the same objective value.

In problem (5.35), the non-convexity is caused by non-convex STARS constraints (5.35b) and (5.35c). To tackle this challenge, we define the auxiliary variables $\tilde{\boldsymbol{\theta}}_i = [\tilde{\beta}_{i,1}e^{j\tilde{\phi}_{i,1}}, \ldots, \tilde{\beta}_{i,N}e^{j\tilde{\phi}_{i,N}}]^T, \forall i\in\{t,r\}$, such that $\tilde{\boldsymbol{\theta}}_i=\boldsymbol{\theta}_i, \forall i\in\{t,r\}$. Then, problem (5.35) can be rewritten as

$$\min_{\mathbf{x}\in\mathcal{X},\boldsymbol{\theta}_t,\boldsymbol{\theta}_r,\tilde{\boldsymbol{\theta}}_t,\tilde{\boldsymbol{\theta}}_r} \quad F(\mathbf{x},\boldsymbol{\theta}_t,\boldsymbol{\theta}_r) \tag{5.36a}$$

$$\text{s.t.} \quad \tilde{\boldsymbol{\theta}}_t=\boldsymbol{\theta}_t, \tilde{\boldsymbol{\theta}}_r=\boldsymbol{\theta}_r, \tag{5.36b}$$

$$\tilde{\beta}_{t,n}^2+\tilde{\beta}_{r,n}^2=1, \forall n\in\mathcal{N}, \tag{5.36c}$$

$$\cos(\tilde{\phi}_{t,n}-\tilde{\phi}_{r,n})=0, \forall n\in\mathcal{N}. \tag{5.36d}$$

In the above problem, there are no constraints imposed on $\boldsymbol{\theta}_t$ and $\boldsymbol{\theta}_r$, except for equality constraint (5.36b). To handle this equality constraint, we exploit the penalty dual decomposition (PDD) framework (Shi & Hong, 2020), where the original problem is converted into the corresponding augmented Lagrangian (AL) problem by moving the equality constraints as a penalty term to the objective function. The AL problem corresponding to (5.36) is given by

$$\min_{\substack{\mathbf{x}\in\mathcal{X},\boldsymbol{\theta}_t,\boldsymbol{\theta}_r,\\ \tilde{\boldsymbol{\theta}}_t,\tilde{\boldsymbol{\theta}}_r}} \quad F(\mathbf{x},\boldsymbol{\theta}_t,\boldsymbol{\theta}_r)+\frac{1}{2\rho}\sum_{i\in\{t,r\}}\|\tilde{\boldsymbol{\theta}}_i-\boldsymbol{\theta}_i+\rho\boldsymbol{\lambda}_i\|^2 \tag{5.37a}$$

$$\text{s.t.} \quad (5.36\text{c}), (5.36\text{d}), \tag{5.37b}$$

where $\rho > 0$ denotes the penalty factor penalizing the violation of constraint (5.36b) and $\boldsymbol{\lambda}_i, \forall i \in \{t, r\}$, denotes the Lagrangian dual variables. As can be observed, when $\rho \to 0$, then the penalty term will be forced to zero, i.e., equality constraint (5.36b) is enforced. It has been proved that updating the primal and dual variables as well as the penalty factor in an alternating manner, a KKT optimal solution can be obtained via PDD under some mild conditions, such as the *Robinson's condition* or the *Mangasarian-Fromovitz constraint qualification (MFCQ)* condition (Shi & Hong, 2020). Thus, in the following, we focus on solving problem (5.37) by invoking the block successive upper-bound minimization (BSUM) or block coordinate descent (BCD) methods,[2] where we divide the optimization variables into two blocks, namely $\{\mathbf{x}, \boldsymbol{\theta}_t, \boldsymbol{\theta}_r\}$ and $\{\tilde{\boldsymbol{\theta}}_t, \tilde{\boldsymbol{\theta}}_r\}$.

5.4.1.1 Subproblem with Respect to $\{\mathbf{x}, \boldsymbol{\theta}_t, \boldsymbol{\theta}_r\}$

Note that introducing the penalty term has no influence on the convexity of the objective function. Thus, the subproblem with respect to $\{\mathbf{x}, \boldsymbol{\theta}_t, \boldsymbol{\theta}_r\}$ is convex and given by

$$\min_{\mathbf{x}\in\mathcal{X},\boldsymbol{\theta}_t,\boldsymbol{\theta}_r} \; F(\mathbf{x}, \boldsymbol{\theta}_t, \boldsymbol{\theta}_r) + \frac{1}{2\rho} \sum_{i\in\{t,r\}} \|\tilde{\boldsymbol{\theta}}_i - \boldsymbol{\theta}_i + \rho\boldsymbol{\lambda}_i\|^2. \tag{5.38}$$

As a consequence, the optimal solution to the above problem can be efficiently obtained.

5.4.1.2 Subproblem with Respect to $\{\tilde{\boldsymbol{\theta}}_t, \tilde{\boldsymbol{\theta}}_r\}$

Since variables $\{\tilde{\boldsymbol{\theta}}_t, \tilde{\boldsymbol{\theta}}_r\}$ only appear in the constraints and in the penalty term of the objective function, the resulting problem is given by

$$\min_{\tilde{\boldsymbol{\theta}}_t,\tilde{\boldsymbol{\theta}}_r} \sum_{i\in\{t,r\}} \|\tilde{\boldsymbol{\theta}}_i + \boldsymbol{\vartheta}_i\|^2 \tag{5.39a}$$

$$\text{s.t.} \quad \tilde{\beta}_{t,n}^2 + \tilde{\beta}_{r,n}^2 = 1, \forall n \in \mathcal{N}, \tag{5.39b}$$

$$\cos(\tilde{\phi}_{t,n} - \tilde{\phi}_{r,n}) = 0, \forall n \in \mathcal{N}, \tag{5.39c}$$

where $\boldsymbol{\vartheta}_i = -\boldsymbol{\theta}_i + \rho\boldsymbol{\lambda}_i, \forall i \in \{t, r\}$. Although the constraints of the above optimization problem are non-convex, we show that a high-quality solution to this problem can be obtained with low complexity by alternately optimizing the

[2] For BSUM, a locally tight upper bound of the original objective function is optimized in each block. BSUM reduces to BCD when the upper bound is replaced by the original objective function itself (Shi et al., 2020).

amplitudes and phase shifts. Firstly, the objective function of problem (5.39) can be reformulated as follows:

$$\begin{aligned}\sum_{i\in\{t,r\}} \|\tilde{\boldsymbol{\theta}}_i + \boldsymbol{\vartheta}_i\|^2 &= \sum_{i\in\{t,r\}} (\tilde{\boldsymbol{\theta}}_i^H\tilde{\boldsymbol{\theta}}_i + \boldsymbol{\vartheta}_i^H\boldsymbol{\vartheta}_i + 2\mathrm{Re}(\boldsymbol{\vartheta}_i^H\tilde{\boldsymbol{\theta}}_i)) \\ &= \sum_{i\in\{t,r\}}\sum_{n\in\mathcal{N}} \tilde{\beta}_{i,n}^2 + \sum_{i\in\{t,r\}} \boldsymbol{\vartheta}_i^H\boldsymbol{\vartheta}_i + \sum_{i\in\{t,r\}} 2\mathrm{Re}(\boldsymbol{\vartheta}_i^H\tilde{\boldsymbol{\theta}}_i) \\ &\overset{(a)}{=} N + \sum_{i\in\{t,r\}} \boldsymbol{\vartheta}_i^H\boldsymbol{\vartheta}_i + \sum_{i\in\{t,r\}} 2\mathrm{Re}(\boldsymbol{\vartheta}_i^H\tilde{\boldsymbol{\theta}}_i),\end{aligned} \tag{5.40}$$

where (a) is due to constraint $\tilde{\beta}_{t,n}^2 + \tilde{\beta}_{r,n}^2 = 1$. In the objective function, only the term $\sum_{i\in\{t,r\}} 2\mathrm{Re}(\boldsymbol{\vartheta}_i^H\tilde{\boldsymbol{\theta}}_i)$ involves the optimization variables, while the other terms are constants. Then, we decompose $\tilde{\boldsymbol{\theta}}_i$ to amplitude vector $\tilde{\boldsymbol{\beta}}_i = [\beta_{i,1}, \ldots, \beta_{i,N}]^T$ and phase-shift vector $\tilde{\boldsymbol{\psi}}_i = [e^{j\phi_{i,1}}, \ldots, e^{j\phi_{i,N}}]^T$, i.e.,

$$\tilde{\boldsymbol{\theta}}_i = \mathrm{diag}(\tilde{\boldsymbol{\beta}}_i)\tilde{\boldsymbol{\psi}}_i = \mathrm{diag}(\tilde{\boldsymbol{\psi}}_i)\tilde{\boldsymbol{\beta}}_i, \forall i \in \{t, r\}. \tag{5.41}$$

Consequently, problem (5.39) can be rewritten as follows:

$$\min_{\tilde{\boldsymbol{\beta}}_t,\tilde{\boldsymbol{\psi}}_t,\tilde{\boldsymbol{\beta}}_r,\tilde{\boldsymbol{\psi}}_r} \sum_{i\in\{t,r\}} \mathrm{Re}(\boldsymbol{\vartheta}_i \mathrm{diag}(\tilde{\boldsymbol{\beta}}_i)\tilde{\boldsymbol{\psi}}_i) \tag{5.42a}$$

$$\text{s.t.} \quad \tilde{\beta}_{t,n}^2 + \tilde{\beta}_{r,n}^2 = 1, 0 \le \tilde{\beta}_{t,n}, \tilde{\beta}_{r,n} \le 1, \forall n \in \mathcal{N}, \tag{5.42b}$$

$$\cos(\tilde{\phi}_{t,n} - \tilde{\phi}_{r,n}) = 0, \forall n \in \mathcal{N}. \tag{5.42c}$$

Here, we introduce the constraint $0 \le \tilde{\beta}_{t,n}, \tilde{\beta}_{r,n} \le 1$ back to ensure that all the phase shifts are collected in $\tilde{\phi}_{t,n}$ and $\tilde{\phi}_{t,n}$ during the following optimization process, which does not affect the optimal objective value based on the previous analysis. To solve the above optimization problem, we introduce the following two propositions.

Proposition 5.2 (Closed-Form Optimal Solution for Phase Shifts for Given Amplitudes) *Define* $\tilde{\boldsymbol{\vartheta}}_i = \mathrm{diag}(\tilde{\boldsymbol{\beta}}_i^H)\boldsymbol{\vartheta}_i = [\tilde{\vartheta}_{i,1}, \ldots, \tilde{\vartheta}_{i,N}]^T$, $\varphi_n^+ = \tilde{\vartheta}_{t,n}^* + j\tilde{\vartheta}_{r,n}^*$, *and* $\varphi_n^- = \tilde{\vartheta}_{t,n}^* - j\tilde{\vartheta}_{r,n}^*$. *Then, for any given* $\tilde{\boldsymbol{\beta}}_t$ *and* $\tilde{\boldsymbol{\beta}}_r$, *the optimal solutions for the elements of* $\tilde{\boldsymbol{\psi}}_t$ *and* $\tilde{\boldsymbol{\psi}}_r$ *are given by*

$$(\tilde{\psi}_{t,n}^\star, \tilde{\psi}_{r,n}^\star) = \underset{(\tilde{\psi}_{t,n},\tilde{\psi}_{r,n})\in\chi_\psi^n}{\mathrm{argmin}} \mathrm{Re}(\tilde{\vartheta}_{t,n}^*\tilde{\psi}_{t,n}) + \mathrm{Re}(\tilde{\vartheta}_{r,n}^*\tilde{\psi}_{r,n}), \tag{5.43}$$

where χ_ψ^n *denotes a set of a pair of closed-form solutions:*

$$\chi_\psi^n = \left\{(e^{j(\pi-\angle\varphi_n^+)}, e^{j(\frac{3}{2}\pi-\angle\varphi_n^+)}), (e^{j(\pi-\angle\varphi_n^-)}, e^{j(\frac{1}{2}\pi-\angle\varphi_n^-)})\right\}. \tag{5.44}$$

Algorithm 3 BSUM/BCD algorithm for solving problem (5.37)

1: Initialize the optimization variables.
2: **repeat**
3: Update $\{\mathbf{x}, \tilde{\boldsymbol{\theta}}_t, \tilde{\boldsymbol{\theta}}_r\}$ by solving problem (5.38).
4: Update $\{\tilde{\boldsymbol{\psi}}_t, \tilde{\boldsymbol{\psi}}_r\}$ by (5.43).
5: Update $\{\tilde{\boldsymbol{\beta}}_t, \tilde{\boldsymbol{\beta}}_r\}$ by (5.45).
6: **until** Convergence.

Proof See Appendix D. □

Proposition 5.3 (Closed-Form Optimal Solution for Amplitudes for Given Phase Shifts) *Define* $\breve{\boldsymbol{\vartheta}}_i = \mathrm{diag}(\tilde{\boldsymbol{\psi}}_i^H)\boldsymbol{\vartheta}_i = [\breve{\vartheta}_{i,1}, \ldots, \breve{\vartheta}_{i,N}]^T$, $a_n = |\breve{\vartheta}_t^*| \cos(\angle\breve{\vartheta}_t^*)$, $b_n = |\breve{\vartheta}_r^*| \sin(\angle\breve{\vartheta}_r^*)$, $\xi_n = \mathrm{sgn}(b_n) \arccos(\frac{a_n}{\sqrt{a_n^2+b_n^2}})$. *Then, for any given* $\tilde{\boldsymbol{\psi}}_t$ *and* $\tilde{\boldsymbol{\psi}}_r$, *the optimal solutions for the elements of* $\tilde{\boldsymbol{\beta}}_t$ *and* $\tilde{\boldsymbol{\beta}}_r$ *are given by*

$$\tilde{\beta}_{t,n}^\star = \sin\omega_n, \quad \tilde{\beta}_{r,n}^\star = \cos\omega_n, \tag{5.45}$$

where

$$\omega_n = \begin{cases} -\frac{1}{2}\pi - \xi_n, & \text{if } \xi_n \in [-\pi, -\frac{1}{2}\pi), \\ 0, & \text{if } \xi_n \in [-\frac{1}{2}\pi, \frac{1}{4}\pi), \\ \frac{1}{2}\pi, & \text{otherwise}. \end{cases} \tag{5.46}$$

Proof See Appendix E. □

According to Propositions 5.2 and 5.3, we can further divide block $\{\tilde{\boldsymbol{\theta}}_t, \tilde{\boldsymbol{\theta}}_r\}$ into two sub-blocks, namely $\{\tilde{\boldsymbol{\psi}}_t, \tilde{\boldsymbol{\psi}}_r\}$ and $\{\tilde{\boldsymbol{\beta}}_t, \tilde{\boldsymbol{\beta}}_r\}$. Then, the overall BSUM/BCD algorithm to solve problem (5.37) is summarized in Algorithm 3. Since the optimal solution is obtained in each step, the convergence of Algorithm 3 is guaranteed (Razaviyayn et al., 2013). The complexities of updating $\{\tilde{\boldsymbol{\psi}}_t, \tilde{\boldsymbol{\psi}}_r\}$ and $\{\tilde{\boldsymbol{\beta}}_t, \tilde{\boldsymbol{\beta}}_r\}$ are $O(4N)$ and $O(2N)$, respectively. Moreover, the complexity of updating $\{\mathbf{x}, \boldsymbol{\theta}_t, \boldsymbol{\theta}_r\}$ is determined by the exact form of problem (5.38) and the methods used to solve it.

5.4.2 System Model and Problem Formulation

To verify the effectiveness of the proposed general optimization framework, we use a case study, where we maximize the throughput in a narrowband STARS-aided communication system.

5.4.2.1 System Model

Consider an M-antenna BS, an N-element STARS, and K single-antenna users, whose indices are collected in $\mathcal{K}$. Without loss of generality, we assume that the users in subset $\mathcal{K}_t = \{1, \ldots, K_0\}$ are located on the transmission side, and the users in subset $\mathcal{K}_r = \{K_0 + 1, \ldots, K\}$ are located on the reflection side. The direct BS-user channels are assumed to be blocked. Thus, the received signal at user k, $\forall k \in \mathcal{K}_i, \forall i \in \{t, r\}$, is given by

$$y_k = \mathbf{h}_k^H \mathbf{\Theta}_i \mathbf{G} \sum_{\ell \in \mathcal{K}} \mathbf{w}_\ell s_\ell + n_k, \tag{5.47}$$

where $\mathbf{h}_k \in \mathbb{C}^{N \times 1}$ denotes the STARS-user-k channel, $\mathbf{G} \in \mathbb{C}^{N \times M}$ denotes the BS-STARS channel, $\mathbf{\Theta}_i = \mathrm{diag}(\boldsymbol{\theta}_i)$ denotes the TCs/RCs of the STARS, $\mathbf{w}_\ell \in \mathbb{C}^{M \times 1}$ denotes the active beamforming vector at the BS for delivering information symbol $s_\ell \in \mathbb{C}$ to user ℓ, and $n_k \sim \mathcal{CN}(0, \sigma_k^2)$ denotes complex Gaussian noise with power σ_k^2. Then, the SINR for decoding s_k at user k, $\forall k \in \mathcal{K}_i, \forall i \in \{t, r\}$, is given by

$$\gamma_k = \frac{|\mathbf{h}_k^H \mathbf{\Theta}_i \mathbf{G} \mathbf{w}_k|^2}{\sum_{\ell \in \mathcal{K} \setminus k} |\mathbf{h}_k^H \mathbf{\Theta}_i \mathbf{G} \mathbf{w}_\ell|^2 + \sigma_k^2}. \tag{5.48}$$

The corresponding achievable rate is $R_k = \log_2(1 + \gamma_k)$.

5.4.2.2 Problem Formulation

We aim to maximize the throughput of the considered STARS-aided system subject to a transmit power constraint and coupled phase-shift constraints. The corresponding optimization problem can be formulated as follows:

$$\max_{\mathbf{W}, \boldsymbol{\theta}_t, \boldsymbol{\theta}_r} \quad \sum_{k \in \mathcal{K}} R_k \tag{5.49a}$$

$$\text{s.t.} \quad \mathrm{tr}(\mathbf{W}\mathbf{W}^H) \leq P_t, \tag{5.49b}$$

$$\beta_{t,n}^2 + \beta_{r,n}^2 = 1, \forall n \in \mathcal{N}, \tag{5.49c}$$

$$\cos(\phi_{t,n} - \phi_{r,n}) = 0, \forall n \in \mathcal{N}, \tag{5.49d}$$

where $\mathbf{W} = [\mathbf{w}_1, \ldots, \mathbf{w}_K]$ and P_t denotes the BS transmit power. We note that existing methods for solving throughput maximization problems cannot be directly applied to problem (5.49) due to the coupled STARS phase-shift and amplitude constraints. In the following section, we show that the proposed general optimization framework can be used to effectively solve problem (5.49).

5.4.3 Solution Using the Proposed Framework

Note that the objective function of problem (5.49) is non-convex with respect to $\{\mathbf{W}, \boldsymbol{\theta}_t, \boldsymbol{\theta}_r\}$. In order to employ the proposed optimization framework, we first transform the throughput maximization problem into an equivalent weighted mean square error (MSE) minimization problem applying the well-known weighted minimum mean square error (WMMSE) method (Christensen et al., 2008) as follows:

$$\max_{\boldsymbol{\varpi}, \boldsymbol{\upsilon}, \mathbf{W}, \boldsymbol{\theta}_t, \boldsymbol{\theta}_r} \sum_{k \in \mathcal{K}} \varpi_k e_k \tag{5.50a}$$

$$\text{s.t.} \quad (5.49\text{b}) - -(5.49\text{d}). \tag{5.50b}$$

Here, $\boldsymbol{\varpi} = [\varpi_1, \ldots, \varpi_K]^T$ denotes the vector of weights, and $e_k, \forall k \in \mathcal{K}_i, \forall i \in \{t, r\}$, denotes the MSE as follows:

$$e_k = |\upsilon_k|^2 \Big(\sum_{\ell \in \mathcal{K}} |\boldsymbol{\theta}_i^T \text{diag}(\mathbf{h}_k^H) \mathbf{G} \mathbf{w}_\ell|^2 + \sigma_k^2 \Big) - 2\text{Re}\{\upsilon_k^* \boldsymbol{\theta}_i^T \text{diag}(\mathbf{h}_k^H) \mathbf{G} \mathbf{w}_k\} + 1, \tag{5.51}$$

where $\boldsymbol{\upsilon} = [\upsilon_1, \ldots, \upsilon_K]^T$ are auxiliary variables. According to Christensen et al. (2008), it can be proved that if $\{\boldsymbol{\varpi}, \boldsymbol{\upsilon}, \mathbf{W}, \boldsymbol{\theta}_t, \boldsymbol{\theta}_r\}$ is a KKT optimal solution to problem (5.50), $\{\mathbf{W}, \boldsymbol{\theta}_t, \boldsymbol{\theta}_r\}$ is also a KKT optimal solution to problem (5.49). In problem (5.50), the objective function is block-wise convex with respect to $\{\boldsymbol{\varpi}, \boldsymbol{\upsilon}\}$, $\mathbf{W}$, and $\{\boldsymbol{\theta}_t, \boldsymbol{\theta}_r\}$. Moreover, the feasible sets of $\{\boldsymbol{\varpi}, \boldsymbol{\upsilon}\}$ and $\mathbf{W}$ are also convex. Thus, we have transformed the throughput maximization problem into the canonical form of problem (5.35), where the corresponding optimization variables, feasible set, and objective function are given by $\mathbf{x} = \{\boldsymbol{\varpi}, \boldsymbol{\upsilon}, \mathbf{W}\}$, $\mathcal{X} = \big\{(\boldsymbol{\varpi}, \boldsymbol{\upsilon}, \mathbf{W}) | \text{tr}(\mathbf{W}\mathbf{W}^H) \le P_t \big\}$, and $F(\mathbf{x}, \boldsymbol{\theta}_t, \boldsymbol{\theta}_r) = \sum_{k \in \mathcal{K}} \varpi_k e_k$, respectively. Therefore, we can employ the proposed framework to solve problem (5.49).

By defining $\tilde{\boldsymbol{\theta}}_t = \boldsymbol{\theta}_t$ and $\tilde{\boldsymbol{\theta}}_r = \boldsymbol{\theta}_r$, problem (5.50) can be equivalently reformulated as follows:

$$\min_{\substack{\boldsymbol{\varpi}, \boldsymbol{\upsilon}, \mathbf{W}, \boldsymbol{\theta}_t, \boldsymbol{\theta}_r \\ \tilde{\boldsymbol{\theta}}_t, \tilde{\boldsymbol{\theta}}_r}} \sum_{k \in \mathcal{K}} \varpi_k e_k \tag{5.52a}$$

$$\text{s.t.} \quad \tilde{\boldsymbol{\theta}}_t = \boldsymbol{\theta}_t, \tilde{\boldsymbol{\theta}}_r = \boldsymbol{\theta}_r \tag{5.52b}$$

$$\text{tr}(\mathbf{W}\mathbf{W}^H) \le P_t, \tag{5.52c}$$

$$\tilde{\beta}_{t,n}^2 + \tilde{\beta}_{r,n}^2 = 1, \forall n \in \mathcal{N}, \tag{5.52d}$$

$$\cos(\tilde{\phi}_{t,n} - \tilde{\phi}_{r,n}) = 0, \forall n \in \mathcal{N}. \tag{5.52e}$$

By moving equality constraint (5.52b) via a penalty term to the objective function, the following problem is obtained:

$$\min_{\substack{\varpi,\upsilon,\mathbf{W},\boldsymbol{\theta}_t,\boldsymbol{\theta}_r\\ \tilde{\boldsymbol{\theta}}_t,\tilde{\boldsymbol{\theta}}_r}} \sum_{k\in\mathcal{K}} \varpi_k e_k + \frac{1}{2\rho}\sum_{i\in\{t,r\}} \|\tilde{\boldsymbol{\theta}}_i - \boldsymbol{\theta}_i + \rho\boldsymbol{\lambda}_i\|^2 \tag{5.53a}$$

$$\text{s.t.} \quad \text{(5.52c)–(5.52e)}. \tag{5.53b}$$

Then, the above problem can be solved via BCD by alternately optimizing the blocks $\{\varpi, \upsilon\}$, $\mathbf{W}$, $\{\boldsymbol{\theta}_t, \boldsymbol{\theta}_r\}$, $\{\tilde{\boldsymbol{\psi}}_t, \tilde{\boldsymbol{\psi}}_r\}$, and $\{\tilde{\boldsymbol{\beta}}_t, \tilde{\boldsymbol{\beta}}_r\}$.

5.4.3.1 Subproblem with Respect to $\{\varpi, \upsilon\}$

The optimal ϖ_k and $\upsilon_k, \forall k \in \mathcal{K}_i, \forall i \in \{t, r\}$, of this subproblem are given by Christensen et al. (2008)

$$\varpi_k = 1 + \gamma_k, \tag{5.54}$$

$$\upsilon_k = \frac{\boldsymbol{\theta}_i^T \operatorname{diag}(\mathbf{h}_k^H)\mathbf{G}\mathbf{w}_k}{\sum_{\ell\in\mathcal{K}} |\boldsymbol{\theta}_i^T \operatorname{diag}(\mathbf{h}_k^H)\mathbf{G}\mathbf{w}_\ell|^2 + \sigma_k^2}. \tag{5.55}$$

5.4.3.2 Subproblems with Respect to W and $\{\boldsymbol{\theta}_t, \boldsymbol{\theta}_r\}$

Note that the objective function of (5.52) is convex with respect to $\mathbf{W}$ and $\{\boldsymbol{\theta}_t, \boldsymbol{\theta}_r\}$, respectively. Thus, the corresponding optimal solution can be efficiently obtained using existing optimization toolboxes, such as CVX (Grant & Boyd, 2014).

5.4.3.3 Subproblems with Respect to $\{\tilde{\boldsymbol{\psi}}_t, \tilde{\boldsymbol{\psi}}_r\}$ and $\{\tilde{\boldsymbol{\beta}}_t, \tilde{\boldsymbol{\beta}}_r\}$

The optimal $\{\tilde{\boldsymbol{\psi}}_t, \tilde{\boldsymbol{\psi}}_r\}$ and $\{\tilde{\boldsymbol{\beta}}_t, \tilde{\boldsymbol{\beta}}_r\}$ can be directly obtained based on Propositions 5.2 and 5.3.

Finally, the dual variables $\{\boldsymbol{\lambda}_t, \boldsymbol{\lambda}_r\}$ and penalty factor ρ can be updated following the PDD framework. As a consequence, the overall algorithm for solving problem (5.52) is summarized in Algorithm 4, where $\delta = \max\{\|\tilde{\boldsymbol{\theta}}_t - \boldsymbol{\theta}_t\|_\infty, \|\tilde{\boldsymbol{\theta}}_r - \boldsymbol{\theta}_r\|_\infty\}$ denotes the constraint violation function. If the MFCQ condition is satisfied, the PDD framework can obtain a KKT optimal solution to problem (5.52), c.f. Shi and Hong (2020). Thus, we show that the MFCQ condition indeed holds for problem (5.52).

Proposition 5.4 *MFCQ holds for problem* (5.52) *at any feasible point* $\{\mathbf{W}, \boldsymbol{\theta}_t, \boldsymbol{\theta}_r\}$ *with* $\mathbf{W} \neq \mathbf{0}$.

Proof See Appendix F. □

Algorithm 4 PDD-based algorithm for solving problem (5.52)

Initialize the optimization variables, and $0 < c < 1$.
repeat
  repeat
    Update $\{\boldsymbol{\varpi}, \boldsymbol{\upsilon}\}$ by (5.54) and (5.55).
    Update $\mathbf{W}$ by solving (5.53) for $\mathbf{W}$.
    Update $\{\boldsymbol{\theta}_t, \boldsymbol{\theta}_r\}$ by solving (5.53) for $\{\boldsymbol{\theta}_t, \boldsymbol{\theta}_r\}$.
    Update $\{\tilde{\boldsymbol{\psi}}_t, \tilde{\boldsymbol{\psi}}_r\}$ by (5.43).
    Update $\{\tilde{\boldsymbol{\beta}}_t, \tilde{\boldsymbol{\beta}}_r\}$ by (5.45).
  until convergence
  if $\delta \leq \eta$ **then**
    set $\boldsymbol{\lambda}_i = \boldsymbol{\lambda}_i + \frac{1}{\rho}(\tilde{\boldsymbol{\theta}}_i - \boldsymbol{\theta}_i), \forall i \in \{t, r\}$.
  else
    set $\rho = c\rho$.
  end if
  Set $\eta = 0.9\delta$.
until δ falls below a predefined threshold.

According to Proposition 5.4, we can conclude that a KKT optimal solution to problem (5.52) can be obtained with Algorithm 4, which is also a KKT optimal solution to the original problem (5.49).

Remark 5.1 This case study reveals that once the original problem is transformed into a form for which (1) the objective function is convex or block-wise convex, and (2) the feasible set is convex or block-wise convex except for constraints (5.35b) and (5.35c), the proposed optimization framework can be directly used. Typically, such a transformation can be achieved by existing methods such as WMMSE, majorization-minimization, and SCA. In most cases, if the transformed problem satisfies the mild MFCQ condition, the KKT optimal solution can be obtained when PDD is employed to update the dual variables and the penalty factor in the proposed framework (Shi & Hong, 2020). Otherwise, at least the convergence of the proposed framework can be guaranteed.

5.4.4 Numerical Results

In this section, simulation results are provided to verify the effectiveness of the proposed optimization framework. Here, we assume that the BS with $M = 8$ antennas is 50 meters, under an angle of 20°, away from the STARS. The users are located on half-circles centered at the STARS with a radius of 3 meters. We also assume that half of the users are located on the transmission side and the rest are located on the reflection side. The channels are modeled as Rician fading channels with a Rician factor of 3 dB and a path loss exponent of 2.2. The path loss at the reference distance of 1 meter is set to 30 dB. The transmit power of the BS and the noise power at the users are set to 20 dBm and −110 dBm, respectively.

Fig. 5.9 The convergence behavior of Algorithm 4. (**a**) Throughput for different values of K. (**b**) Phase-shift difference for $K = 6$

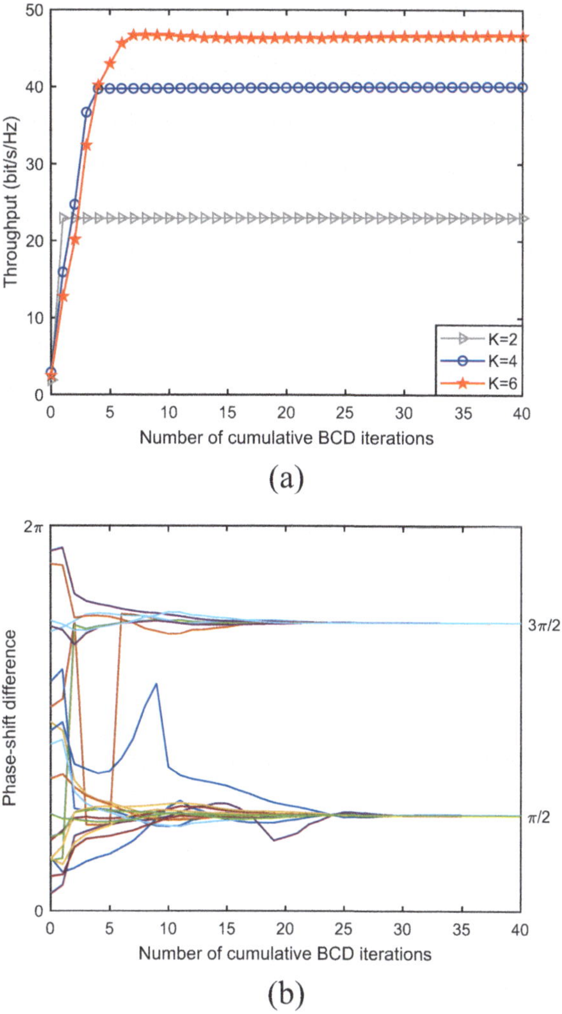

In Fig. 5.9, we first study the convergence of the proposed Algorithm 4 for $N = 20$. As can be seen from Fig. 5.9a, the throughput rapidly converges to a stationary value for all considered values of K. Moreover, Fig. 5.9b shows the absolute phase-shift difference $|\phi_{t,n} - \phi_{r,n}|$ for all 20 elements. As can be observed, the phase-shift differences converge to $\frac{\pi}{2}$ or $\frac{3\pi}{2}$, i.e., $\cos(\phi_{t,n} - \phi_{r,n}) = 0$.

Next, we consider the following benchmark schemes for performance comparison. (1) **STARS, Coupled, AO** (Niu & Liang, 9551): In this scheme, AO is exploited, where the coefficients of one side are optimized subject to the coupled phase-shift constraints by fixing the coefficients of the other side. (2) **STARS, PS-PSC, T** (Xu et al., 2022): This refers to the heuristic primary-secondary phase-shift configuration (PS-PSC) scheme, where the transmission side is primary for STARS.

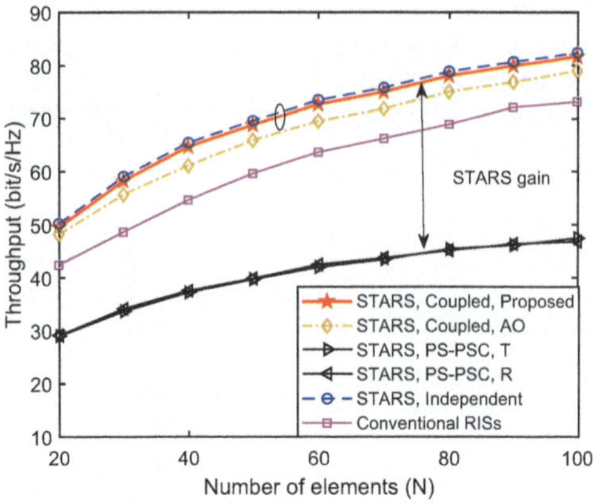

Fig. 5.10 Throughput versus the number of elements N

The STARS coefficients are obtained by fixing the optimal TCs and adjusting RCs such that the coupled phase-shift constraints are satisfied. (3) **STARS, PS-PSC, R** (Xu et al., 2022): This scheme is similar to the previous scheme, but the reflection side is primary. (4) **STARS, Independent** (Mu et al., 2022): In this scheme, the phase-shift coefficients of the STARS can be independently adjusted. (5) **Conventional RIS** (Huang et al., 2019): In this scheme, there are two $\frac{N}{2}$-element reflect-only and transmit-only RISs deployed adjacent to each other.

In Fig. 5.10, we show the throughput versus the number of elements N for different STARS optimization schemes when $K = 6$. The results are obtained by averaging over 100 random channel realizations. As can be observed, the proposed PDD-based optimization framework significantly outperforms the AO-based algorithm and the heuristic PS-PSC schemes for coupled phase-shift STARSs. Moreover, Fig. 5.10 also reveals that the coupled phase-shift model achieves almost the same performance as the independent one and achieves a significant performance gain over the conventional RIS.

5.5 Summary

This chapter introduces three operating protocols for employing STARS in wireless communications and discusses the STARS beamforming design from the independent phase-shift model to the coupled phase-shift model.

References

Ben-Tal, A., & Zibulevsky, M. (1997). Penalty/barrier multiplier methods for convex programming problems. *SIAM Journal on Optimization, 7*, 347–366.

Christensen, S. S., Agarwal, R., De Carvalho, E., & Cioffi, J. M. (2008). Weighted sum-rate maximization using weighted MMSE for MIMO-BC beamforming design. *IEEE Transactions on Wireless Communications, 7*, 4792–4799.

Dinh, Q. T., & Diehl, M. (2010). Local convergence of sequential convex programming for non-convex optimization. In: *Recent advances in optimization and its applications in engineering*. Springer.

Grant, M., & Boyd, S. (2014). CVX: Matlab software for disciplined convex programming, version 2.1. http://cvxr.com/cvx

Huang, C., Zappone, A., Alexandropoulos, G. C., Debbah, M., & Yuen, C. (2019). Reconfigurable intelligent surfaces for energy efficiency in wireless communication. *IEEE Transactions on Wireless Communications, 18*, 4157–4170.

Luo, Z., Ma, W., So, A. M., Ye, Y., & Zhang, S. (2010). Semidefinite relaxation of quadratic optimization problems. *IEEE Signal Processing Magazine, 27*, 20–34.

Mu, X., Liu, Y., Guo, L., Lin, J., & Schober, R. (2022). Simultaneously transmitting and reflecting (STAR) RIS aided wireless communications. *IEEE Transactions on Wireless Communications, 21*, 3083–3098.

Niu, H., & Liang, X. (2022). Weighted sum-rate maximization for STAR-RISs-aided networks with coupled phase-shifters. *IEEE Systems Journal*. https://doi.org/10.1109/JSYST.2022.3159551

Nocedal, J., & Wright, S. (2006). *Numerical optimization*. Springer.

Razaviyayn, M., Hong, M., Luo, Z.-Q. (2013). A unified convergence analysis of block successive minimization methods for nonsmooth optimization. *SIAM Journal on Optimization, 23*, 1126–1153.

Shi, Q., & Hong, M. (2020). Penalty dual decomposition method for nonsmooth nonconvex optimization—part I: Algorithms and convergence analysis. *IEEE Transactions on Signal Processing, 68*, 4108–4122.

Shi, Q., Hong, M., Fu, X., & Chang, T.-H. (2020). Penalty dual decomposition method for nonsmooth nonconvex optimization—part II: Applications. *IEEE Transactions on Signal Processing, 68*, 4242–4257.

Wu, Q., & Zhang, R. (2019). Intelligent reflecting surface enhanced wireless network via joint active and passive beamforming. *IEEE Transactions on Wireless Communications, 18*, 5394–5409.

Xu, J., Liu, Y., Mu, X., Schober, R., & Poor, H. V. (2022). STAR-RISs: A correlated T&R phase-shift model and practical phase-shift configuration strategies. *IEEE Journal of Selected Topics in Signal Processing, 16*, 1097–1111.

Zhang, S., & Zhang, R. (2021). Intelligent reflecting surface aided multi-user communication: Capacity region and deployment strategy. *IEEE Transactions on Communications, 69*, 5790–5806.

Chapter 6
Machine Learning Empowered STARS

Abstract In this chapter, we will introduce the application of machine learning (ML) tools in STARS-aided wireless communications. Two novel designs will be discussed for addressing the beamforming design for STARS-aided wireless communications.

6.1 Motivation of Employing Machine Learning for STARS

For existing research contributions utilizing conventional convex optimization-based methods for STARS-aided wireless communications, these developed algorithms may rely on some ideal assumptions (e.g., perfect CSI, ignored STARS configuration overhead, etc.) and sometimes suffer from high computational complexity, thus becoming inefficient in practical implementation. As a remedy, with the rapid development of AI, machine learning (ML) tools have become an efficient enabler for 6G and beyond wireless communication designs. Therefore, ML-empowered STARS-aided wireless communication design has attracted significant research interest in recent years. In the following, we will first discuss the employment of ML tools to address STARS beamforming design under the coupled phase-shift model, where a pair of hybrid reinforcement learning (RL) algorithms, namely the hybrid deep deterministic policy gradient (hybrid DDPG) algorithm as well as the joint DDPG and deep-Q network (DDPG-DQN) based algorithm, are proposed. Next, we will focus on the ML-empowered STARS-aided multicell NOMA communication design, where a tile-based STARS beamforming strategy is proposed with the employment of a distributed learning approach.

Y. Liu et al., *Simultaneously Transmitting and Reflecting Surfaces for Wireless Communications*, Wireless Networks, https://doi.org/10.1007/978-3-031-76848-4_6

6.2 Hybrid Reinforcement Learning Empowered STARS

Note that the efficient STARS beamforming plays a critical role for maximizing the performance gain of STARS-aided wireless communications. However, for STARS with the coupled phase-shift model, the resulting STARS beamforming becomes more challenging due to the coupling between TCs and RCs. This requires a hybrid continuous and discrete control scheme for the phase-shift design. Given the above-mentioned technical barriers, it is a challenge to solve the STARS beamforming problem, especially considering that the existing convex optimization and ML solutions basically only support either continuous or discrete control. Although several hybrid algorithms have been proposed in the field of computer science (Neunert et al., 2020; Delalleau et al., 2019; Li et al., 2021a), they are designed for minuscule action dimensions. For example, only four discrete actions were assumed in Delalleau et al. (2019) since gaming controllers usually have four buttons. However, the possible number of actions can be a^N for the STARS, where a represents the possible number of actions for a single STARS element and N represents the total number of elements employed. Based on the current assumptions, STARS prototypes are likely to have a massive number of elements, which implies that the action dimension of STARSs substantially exceeds the design in existing algorithms. Since it lacks suitable hybrid algorithms for the optimization of STARS-aided wireless communications, in the following, we will propose two efficient hybrid RL algorithms for facilitating the joint active BS and passive STARS beamforming design.

6.2.1 *System Model and Problem Formulation*

6.2.1.1 System Model

Here, we still consider a STARS operating in the ES protocol, which is deployed in the wireless network. Mobile users can be served by the transmitted or reflected signal, respectively, depending on which region they happen to be roaming in. In order to perform joint beamforming to cover both the transmission and reflection sectors, the TC and RC of each STARS element have to be appreciatively integrated, which are denoted by $\beta_{R,n}e^{j\theta_{R,n}}$ and $\beta_{T,n}e^{j\theta_{T,n}}$, $n = 1, 2, \ldots, N$. Here, we focus on the coupled phase-shift STARS. Therefore, for STARS elements, the coupling between the TC's phase shift $\theta_{T,n}$, the RC's phase shift $\theta_{R,n}$, and the amplitude β_n follows a relationship as

$$\beta_n\sqrt{1-\beta_n^2}\cos(\theta_{R,n}-\theta_{T,n}) = 0. \tag{6.1}$$

Thus, for a STARS having N elements, the transmission and reflection matrices have a diagonal structure given by

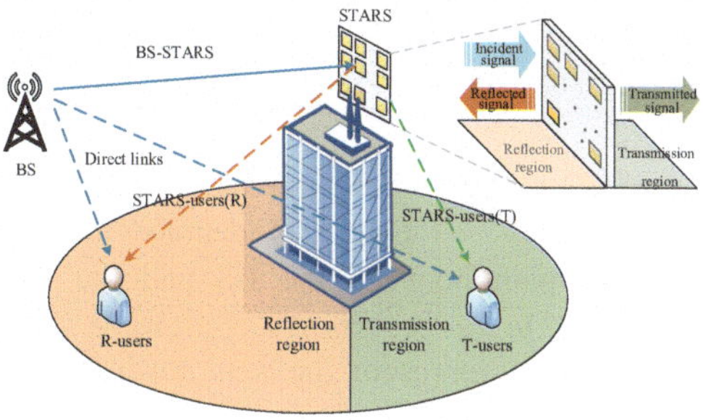

Fig. 6.1 System model of STARS-aided wireless networks

$$\mathbf{\Theta}_R = \mathrm{diag}\left(\beta_1 e^{j\theta_{R,1}}, \beta_2 e^{j\theta_{R,2}}, \cdots, \beta_N e^{j\theta_{R,N}}\right), \tag{6.2}$$

$$\mathbf{\Theta}_T = \mathrm{diag}\left(\sqrt{1-\beta_1^2} e^{j\theta_{T,1}}, \cdots, \sqrt{1-\beta_N^2} e^{j\theta_{T,N}}\right). \tag{6.3}$$

We consider a downlink scenario of Fig. 6.1, where the BS is equipped with M antennas, and the STARS has N elements. There are K randomly roaming users and each has a single antenna. Each of the N STARS elements has the amplitude response $\beta_n, n = 1, 2, \ldots, N$. We denote the locations of the BS, RIS, and users as $(x_b, y_b, z_b)^T$, $(x_r, y_r, z_r)^T$, and $(x_k, y_k, z_k)^T$, respectively. The STARS naturally partitions the users into two groups according to their locations. The users between the BS and the STARS receive direct signals from the BS and reflected signals from the STARS. This set of users in the reflective region of the STARS is denoted by R_u. For simplicity, we term this set of users as R users in the rest of the text. Correspondingly, the users served by direct BS signals and transmitted STARS signals are denoted by T_i. The number of users obeys $K = U + I$, where U and I are the numbers of R users and T users.[1]

In the STARS-aided wireless communication scenario, multiple channels have to be considered, including the BS to STARS channel $\boldsymbol{H}_{b,r} \in \mathbb{C}^{M\times N}$, the direct channel spanning the BS to R and T users $\boldsymbol{H}_{b,R} \in \mathbb{C}^{M\times U}$, $\boldsymbol{H}_{b,T} \in \mathbb{C}^{M\times I}$, and the channel impinging from the STARS to R and T users $\boldsymbol{H}_{r,R} \in \mathbb{C}^{N\times U}$, $\boldsymbol{H}_{r,T} \in \mathbb{C}^{N\times I}$. For each specific user, R_u and T_i, the direct channels, and the $T\&R$ channels can be denoted as $\boldsymbol{h}_{b,u} \in \mathbb{C}^{M\times 1}$, $\boldsymbol{h}_{b,i} \in \mathbb{C}^{M\times 1}$, $\boldsymbol{h}_{r,u} \in \mathbb{C}^{N\times 1}$, and $\boldsymbol{h}_{r,i} \in \mathbb{C}^{N\times 1}$, respectively.

All channels are assumed to follow the quasi-static block-fading model, where the fading coefficient remains constant in each time slot (TS) t. We assume that the channel $\boldsymbol{H}_{b,r}$ has an LoS path and obeys to the Rician distribution since the BS and STARS have an LoS component owing to their selected positions. Thus, upon considering the path loss and the small-scale fading, the Rician channel can be formulated as

[1] For simplicity of equations, the subscript $k = 1, 2, \ldots, K$ refers to any user, user $R_u, u = 1, 2, \ldots, U$ or $T_i, i = 1, 2, \ldots, I$ refers to a R user or T user.

$$\boldsymbol{H}_{b,r,t} = \sqrt{L_{b,r,t}\left(d_{b,r,t}, f_c\right)}\left(\sqrt{\frac{K}{K+1}}\boldsymbol{L}_{b,r,t} + \sqrt{\frac{1}{K+1}}\boldsymbol{G}_{b,r,t}\right), \tag{6.4}$$

where $L_{b,r,t}$ represents the path loss in the power domain, K is the Rician factor, $\mathbf{G}_{b,r,t} \sim CN(0,1)$ denotes the scattered paths, and $\boldsymbol{L}_{b,r,t}$ represents the LoS path between the BS and the RIS. Furthermore, we have $\boldsymbol{L}_{b,r,t} = \boldsymbol{a}_r(\varphi^A, \psi^A)\boldsymbol{a}_b(\varphi^D)^H$, $\boldsymbol{a}_b \in \mathbb{C}^{M\times 1}$, $\boldsymbol{a}_r \in \mathbb{C}^{N\times 1}$, where $\varphi^{A/D} \in [0, 2\pi)$ and $\psi^A \in [-2/\pi, 2/\pi)$ represent the azimuth and elevation angle-of-arrival (AoA)/angle-of-departure (AoD), $\varphi^A = \arcsin(\frac{y_r - y_b}{\sqrt{(x_r-x_b)^2+(y_r-y_b)^2}})$, $\varphi^D = \pi/2 - \varphi^A$, and $\psi^A = \arcsin(\frac{z_r - z_b}{\sqrt{(x_r-x_b)^2+(y_r-y_b)^2}})$.

We assume that the BS has linear array antennas and the STARS RIS has a UPA of reflective elements, where the antenna space and the reflective element space are d_a and d_e. According to Zhang and Zhang (2020), we have

$$a_b[m] = e^{j2\pi(m-1)d_a\sin(\varphi^D)/\lambda}, \tag{6.5}$$

$$a_r[n] = e^{j2\pi(n-1)d_e(\lfloor n/N_x\rfloor\sin(\varphi^A)\sin(\psi^A)+(n-\lfloor n/N_x\rfloor N_x)\sin(\varphi^A)\cos(\psi^A))/\lambda}, \tag{6.6}$$

where N_x denotes the number of reflective elements in each row and $\lfloor\rfloor$ is the floor function.

The path loss L follows the urban propagation model presented in 3GPP specification TR 36.873 (3rd Generation Partnership Project, 2018). For the path with LoS, the path loss can be given by

$$L_{\text{LoS}}\left(d, f_c\right) = 22.0\log_{10} d + 28.0 + 20\log_{10} f_c, \tag{6.7}$$

where d represents the 3D distance between the Tx and the Rx, while f_c is the carrier frequency. For the NLoS propagation, the path loss is given by

$$L_{\text{NLoS}} = \max[L_{\text{LoS}}\left(d, f_c\right), L_{\text{NLoS}}\left(d, f_c\right)], \tag{6.8}$$

$$L_{\text{NLoS}}\left(d, f_c\right) = 36.7\log 10d + 22.7 + 26\log 10 f_c - 0.3(z_r - 1.5). \tag{6.9}$$

On the other hand, due to the random movements of users, LoS propagation may not necessarily be guaranteed, regardless of whether the transmitting side is the BS or the STARS. Therefore, $\boldsymbol{H}_{b,R,t}$, $\boldsymbol{H}_{b,T,t}$, $\boldsymbol{H}_{r,R,t}$, and $\boldsymbol{H}_{r,T,t}$ are assumed to be NLoS channels and follow Rayleigh fading, which can be expressed as $\boldsymbol{H}_{b,R,t} = \sqrt{L_{b,R,t}}\boldsymbol{G}_{b,R,t}$, $\boldsymbol{H}_{b,T,t} = \sqrt{L_{b,T,t}}\boldsymbol{G}_{b,T,t}$, $\boldsymbol{H}_{r,R,t} = \sqrt{L_{r,R,t}}\boldsymbol{G}_{r,R,t}$, and $\boldsymbol{H}_{r,T,t} = \sqrt{L_{r,T,t}}\boldsymbol{G}_{r,T,t}$.

We denote the information sequence and the active beamforming vectors for user k at the BS by $s_{k,t}$ and $\mathbf{w}_{k,t}, \in \mathbb{C}^{M\times K}$. The signal transmitted at TS t can be expressed as

$$\mathbf{x}_{b,t} = \sum_{k=1}^{K} \mathbf{w}_{k,t} s_{k,t}. \tag{6.10}$$

Then, the incident signal at the STARS is given by

$$\mathbf{x}_{r,t} = \boldsymbol{H}_{b,r,t} \sum_{k=K}^{U} \mathbf{w}_{k,t} s_k + n_0, \tag{6.11}$$

where n_0 represents the AWGN, and the received signal of user R^u is given by

$$y_{u,t} = \left(\boldsymbol{h}_{b,u,t} + \boldsymbol{h}_{r,u,t} \boldsymbol{\Theta}_{R,t} \boldsymbol{H}_{b,r,t}\right) \sum_{u=1}^{U} \mathbf{w}_{u,t} s_{u,t} + n_0. \tag{6.12}$$

Correspondingly, the received signal of user T^i can be represented in a similar form as

$$y_{i,t} = \left(\boldsymbol{h}_{b,i,t} + \boldsymbol{h}_{r,i,t} \boldsymbol{\Theta}_{T,t} \boldsymbol{H}_{b,r,t}\right) \sum_{i=1}^{I} \mathbf{w}_{i,t} s_{i,t} + n_0. \tag{6.13}$$

Given the received signal, the SINR of user R^u and T^i is given by

$$\gamma_{u,t} = \frac{\mid \left(\boldsymbol{h}_{b,u,t} + \boldsymbol{h}_{r,u,t} \boldsymbol{\Theta}_{R,t} \boldsymbol{H}_{b,r,t}\right) \mathbf{w}_{u,t} \mid^2}{\mid \left(\boldsymbol{h}_{b,u,t} + \boldsymbol{h}_{r,u,t} \boldsymbol{\Theta}_{R,t} \boldsymbol{H}_{b,r,t}\right) \sum\limits_{k \le K, k \ne u} \mathbf{w}_{k,t} \mid^2 + \sigma^2}, \tag{6.14}$$

$$\gamma_{i,t} = \frac{\mid \left(\boldsymbol{h}_{b,i,t} + \boldsymbol{h}_{r,i,t} \boldsymbol{\Theta}_{T,t} \boldsymbol{H}_{b,r,t}\right) \mathbf{w}_{i,t} \mid^2}{\mid \left(\boldsymbol{h}_{b,i,t} + \boldsymbol{h}_{r,i,t} \boldsymbol{\Theta}_{T,t} \boldsymbol{H}_{b,r,t}\right) \sum\limits_{k \le K, k \ne i} \mathbf{w}_{k,t} \mid^2 + \sigma^2}, \tag{6.15}$$

where σ^2 represents the noise power. Therefore, given a bandwidth B, the achievable data rate of each user is given by

$$R_{k,t} = B \log_2 \left(1 + \gamma_{k,t}\right). \tag{6.16}$$

6.2.1.2 Problem Formulation

We aim to minimize the power consumption of the BS by jointly optimizing the beamforming vector $\mathbf{w}_k$ for the BS and the TCs as well as the RCs of the STARS. Again, the TCs and RCs of the STARS can be represented by the phase shifts

$\boldsymbol{\Theta}_T$, $\boldsymbol{\Theta}_R$, and amplitude coefficient $\boldsymbol{\beta}$. Therefore, the optimization problem can be formulated as

$$\min_{\mathbf{w},\boldsymbol{\Theta}_{T_0},\boldsymbol{\Theta}_R,\boldsymbol{\beta}} \sum_{t=1}^{T}\sum_{k=1}^{K} \| \mathbf{w}_{k,t}^2 \|, \tag{6.17a}$$

$$\text{s.t.} \quad 0 \le \theta_{T,n,t} < 2\pi, \forall n, \forall t, \tag{6.17b}$$

$$0 \le \theta_{R,n,t} < 2\pi, \forall n, \forall t, \tag{6.17c}$$

$$R_{k,t} \ge R_{\mathrm{QoS}}, \forall k, \forall t, \tag{6.17d}$$

$$0 < \beta_{n,t} \le 1, \forall n, \forall t, \tag{6.17e}$$

$$\beta_{n,t}\sqrt{1-\beta_{n,t}^2}\cos(\theta_{R,n,t}-\theta_{T,n,t}) = 0, \tag{6.17f}$$

$$P_{b,t} \le P_{\max}, \tag{6.17g}$$

where constraints (6.17b) and (6.17c) represent the legitimate range of the TC and RC phase shifts. Constraint (6.17d) is a QoS constraint specifically the minimum data rate. Since (STARS) RISs are passive devices, their amplitude response is limited by the conservation of energy, as shown in (6.17e). Constraint (6.17f) characterizes the phase and amplitude relationship of the TCs and RCs. Finally, (6.17g) is the maximum power constraint for the BS.

The challenge of solving our problem is not only owing to the joint consideration of TCs and RCs but also due to the constraint (6.17e). Given the coupling between $\theta_{T,n}$ and $\theta_{R,n}$, the STARS elements n cannot have independent arbitrary TCs and RCs. Furthermore, although the STARS aspires to continuous phase-shift control for both transmission and reflection, one party of them can only have dualistic options. For example, assuming $\beta_n \neq 0$, once the TC is determined as $\theta_{T,n}$, the RC can only select the phase shift from $\{\theta_{T,n} + \frac{\pi}{2}, \theta_{T,n} - \frac{\pi}{2}\}$. Therefore, the coupled phase-shift model of STARSs requires hybrid continuous and discrete control for the transmission and reflection, which motivates us to develop hybrid deep reinforcement learning (DRL) algorithms for solving this challenge.

6.2.2 *Hybrid DDPG Algorithm for STARS*

The deep deterministic policy gradient (DDPG) algorithm was shown to constitute an efficient solution for continuous control problems (Lillicrap et al., 2015). For applying DRL approaches to solve the optimization problem, the transmission period has to obey a Markov decision process (MDP) (Li et al., 2021c). In TS $t \in \mathcal{T}$, by checking the CSI of the current channels, the agent determines the current state $\mathbf{s}_t \in \mathbf{S}$ and decides to carry out the action $\mathbf{a}_t \in \mathbf{A}$, where $\mathbf{S}$ and $\mathbf{A}$ represent the state space and action space. The action refers to a vector storing the active and passive beamforming coefficients at the BS and STARS.

Since passive and active beamforming are jointly considered, the MDP state for each TS t includes the CSI of the BS to STARS, BS to users (both R and T users), and the STARS to users channels, as shown in Fig. 6.1. Thus, $\mathbf{s}_t$ is given by

$$\mathbf{s}_t = \{\boldsymbol{H}_{b,r,t}, \boldsymbol{H}_{b,R,t}, \boldsymbol{H}_{b,T,t}, \boldsymbol{H}_{r,R,t}, \boldsymbol{H}_{r,T,t}\}. \tag{6.18}$$

Once action $\mathbf{a}_t$ is executed, the agent has to determine the reward r_t according to the data rate and power consumption of the transceiver, and then the state would be constrained to $\mathbf{s}_{t+1} \in \mathbf{S}$. Once the above steps are completed, $(\mathbf{s}_t, \mathbf{a}_t, r_t, \mathbf{s}_{t+1})$ would be saved as a Markov transition in the replay buffer for the agent's training.

6.2.2.1 DDPG Training

The objective of the DRL-based agent training is to find the specific action $\mathbf{a}_t$ for each state $\mathbf{s}_t$, which maximizes the expected accumulated reward $\mathbb{E}[\sum_{i=t}^{T} \gamma r_{t+1}]$, where γ represents the discount factor $\gamma \in [0, 1]$. For a DDPG agent, the Q value of action $\mathbf{a}_t$ can be quantified by the Bellman equation of Feriani and Hossain (2021)

$$Q^{\mu}(\mathbf{s}_t, \mathbf{a}_t) = \mathbb{E}\big[r(\mathbf{s}_t, \mathbf{a}_t) + \gamma Q^{\mu}(\mathbf{s}_{t+1}, \mathbf{a}_{t+1})\big], \tag{6.19}$$

where μ represents the action policy function $\mathbf{S} \leftarrow \mathbf{A}$. The training of DRL agents aims to ascertain the optimal action yielding the maximum Q value, as given by

$$Q^{*}(\mathbf{s}_t, \mathbf{a}_t) = \mathbb{E}\big[r(\mathbf{s}_t, \mathbf{a}_t) + \max_{\mathbf{a} \in \mathbf{A}} \gamma Q^{*}(\mathbf{s}_{t+1}, \mathbf{a}_{t+1})\big]. \tag{6.20}$$

We consider a parameterized actor function $\mu(\mathbf{s}|\boldsymbol{\omega}^{\mu})$ and a function approximator associated with a batch of parameters denoted by $\boldsymbol{\omega}^{Q}$. By sampling the aforementioned transition experiences in the memory of the replay buffer, the DRL agent can be trained by minimizing the loss function

$$L(\boldsymbol{\omega}^{Q}) = \frac{1}{e} \sum_{e} [y_t - Q(\mathbf{s}_t, \mathbf{a}_t | \boldsymbol{\omega}_t^{Q})]^2, \tag{6.21}$$

where e is the size of the sampled transitions. However, to avoid oscillations or divergence during the training process, y_t has to be provided by the target network, which has the same structure as the training network, but is associated with a deferred parameter update. Upon denoting the parameters of the target networks by $\boldsymbol{\omega}^{Q'}$, y_t can be expressed as

$$y_t = r_t(\mathbf{s}_t, \mathbf{a}_t) + \gamma Q'[\mathbf{s}_t, \mu'(\mathbf{s}_t | \boldsymbol{\omega}_t^{\mu'}) | \boldsymbol{\omega}_t^{Q'}]. \tag{6.22}$$

According to the principle of the original DDPG algorithm (Lillicrap et al., 2015), the actor network is trained by the policy gradient calculated by the critic network as

$$\nabla_{\omega^{\mu}} J = \frac{1}{e}\sum_{e} \nabla_{\mathbf{a}} Q(\mathbf{s}_e, \mathbf{a}_e|\boldsymbol{\omega}^{Q})|_{\mathbf{s}_e=\mathbf{s}_t,\mathbf{a}_e=\mu(\mathbf{s}_t)} \nabla_{\omega^{\mu}} \mu(\mathbf{s}_e|\boldsymbol{\omega}^{\mu}|_{\mathbf{s}_e=\mathbf{s}_t}). \tag{6.23}$$

6.2.2.2 Continuous-Discrete Actions and Hybrid DDPG

The actions designed for the DRL agent have to contain all optimized variables, resulting in the action space of $\mathbf{a}_t = \{\mathbf{w}, \boldsymbol{\Theta}_T, \boldsymbol{\Theta}_R, \boldsymbol{\beta}\}$. Among these optimization variables, $\mathbf{w}$ and $\boldsymbol{\beta}$ can be handled by a continuous control scheme to achieve precise control. Therefore, we focus our attention on the discussion of $\boldsymbol{\Theta}_T$ and $\boldsymbol{\Theta}_R$. As described above, due to the existence of the constraint (6.17f), the transmission and reflection phase shifts cannot be adjusted independently by the STARS. For element n, assume $\beta_n \neq 0$ and $\theta_{R,n}$ is determined, in order to satisfy the constraint (6.17f). In this context, it is no hard to discover that $\theta_{T,n} = \theta_{R,n} \pm \frac{\pi}{2}$. Therefore, regardless of whether the phase shift $\theta_{R,n}$ or $\theta_{T,n}$ is selected to be continuously controlled, the phase control of the other one is no longer a continuous control problem, but a binary selection problem. Therefore, (6.17a) requires continuous control for $\mathbf{w}, \boldsymbol{\Theta}_R, \boldsymbol{\beta}$ and discrete control for $\boldsymbol{\Theta}_T$, which lead to a hybrid action space associated with continuous and discrete actions. The continuous subspace and the discrete subspace are denoted by $\mathbf{a}^c = (a_1^c, a_2^c \ldots a_n^c \ldots)$ and $\mathbf{a}^d = (a_1^d, a_2^d \ldots a_n^d \ldots)$,[2] respectively.

1. *Action Space for the DDPG Algorithm*: If the DDPG algorithm is applied to solve the optimization problem associated with the hybrid action space, a direct and intuitive approach is to discretize a part of the continuous output of the DDPG algorithm as presented in Fig. 6.2a and Algorithm 5. Including the action noise, a classic action policy is given for the DDPG algorithm by

$$\mathbf{a}_t = \mu(\mathbf{s}_t|\boldsymbol{\omega}_t^{\mu}) + N_{\mathrm{OU}}(0, \xi), \tag{6.24}$$

where $N_{\mathrm{OU}}(0, \xi)$ represents the zero mean Ornstein Uhlenbeck (OU) noise (Abraham et al., 2021) that follows $N_{\mathrm{OU}} \sim OU(0, \xi)$, and ξ is the volatility of the OU noise.

Thus, the continuous action space can be formulated as

$$\mathbf{a}_t^c = \{\mathbf{a}_t^{\mathbf{w}}, \mathbf{a}_t^{\boldsymbol{\Theta}_R}, \mathbf{a}_t^{\boldsymbol{\beta}}\}. \tag{6.25}$$

[2] For the convenience of presentation, we assume that $\boldsymbol{\Theta}_R \subset a^c$ and $\boldsymbol{\Theta}_T \subset a^d$.

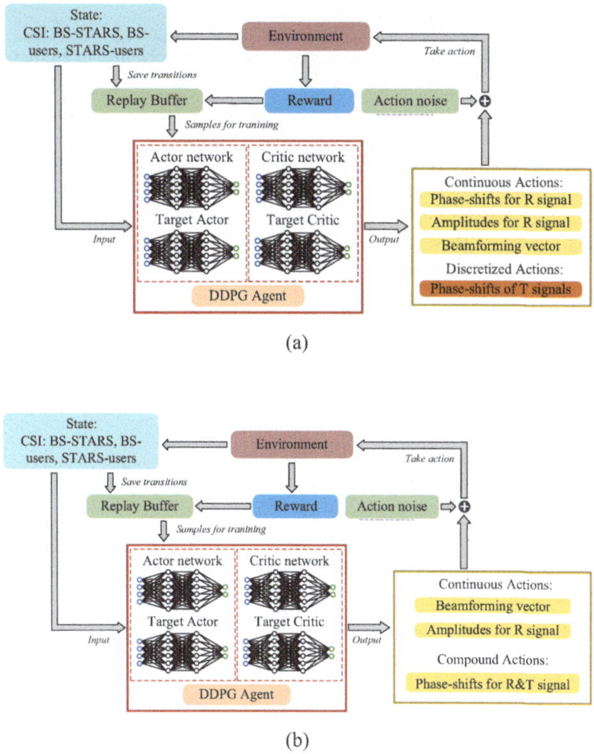

Fig. 6.2 Flow diagrams of the DDPG/hybrid DDPG algorithms. (**a**) Flow diagram of the conventional DDPG algorithm with output discretization. (**b**) Flow diagram of the proposed hybrid DDPG algorithm

Then, the normalized output of the actor network may be decoded into executable actions for our communication environment by following the actions of

$$\mathbf{w} \leftarrow \mathbf{a}_t^{\mathbf{w}}, \tag{6.26}$$

$$\boldsymbol{\Theta}_R \leftarrow \mathbf{a}_t^{\boldsymbol{\Theta}_R}, \tag{6.27}$$

$$\boldsymbol{\beta} \leftarrow \mathbf{a}_t^{\boldsymbol{\beta}}. \tag{6.28}$$

Since the mapping of $\mathbf{a}_t^c$ and the continuous actions are trivial, it is not necessary to discuss them in detail. On the other hand, for a specific STARS element n, $\theta_{T,n}$ can be obtained by the binary discretized $a_{n,t}^d$

$$\theta_{T,n,t} = \begin{cases} \theta_{R,n,t} + \frac{\pi}{2}, & a_{n,t}^d > 0, \\ \theta_{R,n,t} - \frac{\pi}{2}, & a_{n,t}^d \leq 0. \end{cases} \tag{6.29}$$

Algorithm 5 DDPG/Hybrid DDPG algorithm

Initialize the environment and the agent with the actor network $\boldsymbol{\omega}^{\mu}$, critic network $\boldsymbol{\omega}^{Q}$, target actor network $\boldsymbol{\omega}^{\mu'}$, target critic network $\boldsymbol{\omega}^{Q'}$.
for each episode **do**
  Reinitialize the environment to $\mathbf{s}_{t=0}$.
  for each step in $t_0 \leq t \leq T_0$. **do**
    Observe $\mathbf{s}_t$.
    Choose $\mathbf{a}_t$ according to (6.25).
    if DDPG algorithm **then**
      Discretize a part of $\mathbf{a}_t$ into $\mathbf{a}_{t,d}$.
    end if
    if Hybrid DDPG algorithm **then**
      Map $\mathbf{a}_t$ with $\boldsymbol{\Theta}_{R,t}$ and $\boldsymbol{\Theta}_{T,t}$.
    end if
    Execute $\mathbf{a}_t$ in the environment.
    Calculated the reward r_t and observe the next state $\mathbf{s}_{t+1}$.
    Record $e\{\mathbf{s}_t, \mathbf{a}_t, r_t, \mathbf{s}_{t+1}\}$ in memory buffer.
    Random sample a batch of transaction e from the memory buffer.
    Calculate target according to (6.22).
    Train critic network $Q(\mathbf{s}_t, \boldsymbol{\omega}_t^{Q})$ with a gradient descent step (6.21).
    Train actor network $\mu(s_t, \boldsymbol{\omega}_t^{\mu})$ with (6.23).
    Update the target networks $\boldsymbol{\omega}_t^{\mu'} \leftarrow (1-\tau)\boldsymbol{\omega}_t^{\mu'} + \tau\boldsymbol{\omega}_t^{\mu}$, $\boldsymbol{\omega}_t^{Q'} \leftarrow (1-\tau)\boldsymbol{\omega}_t^{Q'} + \tau\boldsymbol{\omega}_t^{Q}$.
    $\mathbf{s}_t \leftarrow \mathbf{s}_{t+1}$.
  end for
end for

2. *Action Space for our Hybrid DDPG Algorithm*: In order to deal with the high-dimensional continuous and discrete action components, specifically for the STARS scenario, we propose a hybrid DDPG algorithm, as shown in Fig. 6.2b and Algorithm 5. By exploiting $\boldsymbol{\Theta}_R$ and $\boldsymbol{\Theta}_T$ have the same dimension for the STARS, we design the codebook between the actor outputs and the phase shifts to achieve hybrid control. The hybrid action policy for the STARS at time t can be expressed by

$$\mu(\mathbf{s}_t|\boldsymbol{\omega}_t^{\mu}) = \mu^c(\mathbf{s}_t|\boldsymbol{\omega}_t^{\mu})\mu^d(\mathbf{s}_t|\boldsymbol{\omega}_t^{\mu}) = \prod_{n=1}^{N} \mu^c(s_{n,t}|\boldsymbol{\omega}_t^{\mu}) \prod_{n=1}^{N} \mu^d(s_{n,t}|\boldsymbol{\omega}_t^{\mu}). \tag{6.30}$$

Specifically, for the STARS element n, the values of $\theta_{R,n,t}$ and $\theta_{T,n,t}$ are compounded and given in a normalized output as

$$\mu(s_{n,t}|\boldsymbol{\omega}_t^{\mu}) = \mu^c(s_{n,t}|\boldsymbol{\omega}_t^{\mu})\mu^d(s_{n,t}|\boldsymbol{\omega}_t^{\mu}). \tag{6.31}$$

In the face of the action noise similar to (6.24), the hybrid action $a_{n,t}^h$ can be obtained from (6.31). Then, the normalized action $a_{n,t}^h$ has to be mapped to $\theta_{R,n,t}$ and $\theta_{R,n,t}$ as

Fig. 6.3 Amplitude response over normalized action output of hybrid DDPG algorithm for STARS ($\beta = 0.5$)

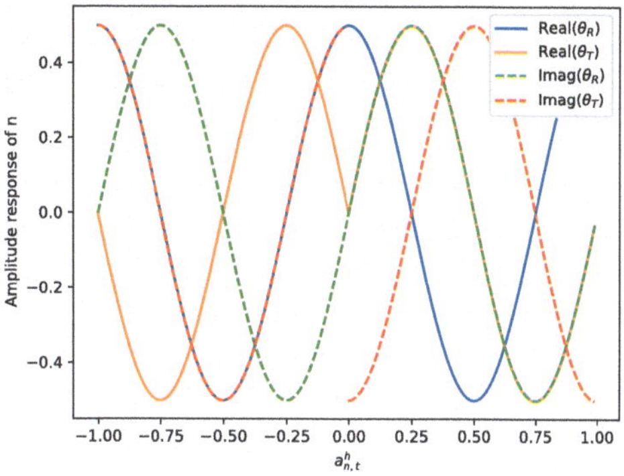

$$\theta_{n,R,t} = 2\pi a^h_{n,t}, \tag{6.32}$$

$$\theta_{n,T,t} = \begin{cases} \theta_{n,R,t} + \frac{\pi}{2}, & a^h_{n,t} > 0, \\ \theta_{n,R,t} - \frac{\pi}{2}, & a^h_{n,t} \leq 0. \end{cases} \tag{6.33}$$

Since the normalized action $a^h_{n,t}$ is in the interval $[-1, 1]$, $\theta_{n,R,t}$ and θ_T have to be mapped with $a^h_{n,t}$. Furthermore, $\theta_{n,R,t}$ vs. $a^h_{n,t}$ is modeled by a periodic linear function and θ_T vs. $a^h_{n,t}$ by a piecewise linear function. The amplitude response $a^h_{n,t}$ of a single STARS element is plotted in Fig. 6.3 for $\beta = 0.5$. Since STARS is generally equipped with a large number of STARS elements, in this case, the hybrid DDPG algorithm has a significantly smaller action dimension than the conventional DDPG algorithm, since we have $|\mathbf{a}^h_t| = \frac{|\mathbf{a}^c_t|+|\mathbf{a}^d_t|}{2}$.

Remark 6.1 In some existing hybrid DRL schemes, the agent can only output a single action for the discrete action space, which is a disadvantage for problems associated with high-dimensional discrete action spaces of multi-antenna or multielement-RIS scenarios. By contrast, the proposed hybrid scheme is eminently suitable for high-dimensional hybrid action spaces, where the discrete action dimension is no larger than the continuous action dimension.

6.2.2.3 Reward Function Design

The plain DDPG algorithm and the hybrid DDPG algorithm have identical reward functions. In order to ensure that the agent meets the users' QoS constraint, each satisfied T and R user contributes a distinguishable positive reward. Meanwhile, to minimize the power consumption, the sum power cost associated with $\mathbf{w}$ results in a negative reward. Thus, the reward function at TS t can be formulated as

$$r_t = \sum_{k=1, R_{k,t} > R_c}^{K} \dot{r} - \sum_{t=1}^{T} \sum_{k=1}^{K} \hat{r} \parallel \mathbf{w}_{k,t}^2 \parallel, \tag{6.34}$$

where the constant coefficient $\dot{r}$ represents the reward gleaned for satisfying the data rate requirement per user, and the coefficient $\hat{r}$ cost (negative reward) for power consumption. It is worth noting that, to guarantee the primary goal of the agent is satisfying the QoS requirement of each and every user rather than saving energy, the coefficients have to satisfy $\dot{r} > \sum_{k=1}^{K} \hat{r} P_{\max}$.

6.2.2.4 Neural Network Structure

In order to ensure accurate fitting, the structure and scale of the (target) actor network and the (target) critic network have to be selected appropriately. The actor networks include the input layer, batch normalization (BN) layer, and activation layer(s) with a "relu" function in turn. Moreover, "tanh" is assigned as the activation function of the output layer. To ensure a valid range for the input values of the output layer, another BN layer has to be employed above the output layer. The critic networks consist of an input layer, BN layer, concatenate layer, and activation layers. The size of the hidden layers has to be determined according to the state and action dimension, which is dependent both on the number of antennas M and on the number of STARS elements N. It is also worth noting that due to the difference in output dimensions between the DDPG scheme and the hybrid DDPG scheme, the appropriate size of the hidden layer may vary, especially when a large number of STARS elements are considered.

6.2.3 *Joint DDPG-DQN Algorithm*

The hybrid DDPG algorithm aims for covering the action space $\mathbf{a}^c$ and $\mathbf{a}^d$ with the aid of a compound action mapping. By contrast, in this section, we explore another promising option, namely that of employing two agents for covering the action spaces $\mathbf{a}^c$ and $\mathbf{a}^d$, respectively. Thus, a DDPG algorithm is harnessed for optimizing $\mathbf{a}^c = \{\mathbf{a}_t^{\mathbf{w}}, \mathbf{a}_t^{\Theta_R}, \mathbf{a}_t^{\beta}\}$, and a DQN algorithm is responsible for giving $\mathbf{a}^d = \mathbf{a}_t^{\Theta_T}$. DDPG and DQN algorithms can be regarded either as a joint agent or as a pair of cooperative agents. For the convenience of the presentation, we regard the DDPG agent and the DQN agent of the joint DDPG-DQN algorithm as a pair of components, but they have to be installed on the same device in practice.

6.2.3.1 MDP for Joint DDPG-DQN

The DDPG and DQN algorithms are capable of optimizing independent continuous and discrete problems (Arulkumaran et al., 2017). Nonetheless, how to harness them in the interest of joint optimization results is an open conundrum. If two agents adopt a parallel relationship at $\mathbf{s}_t$ to output $\mathbf{a}_t^c$ and $\mathbf{a}_t^d$, then according to the classic MDP model, the execution of $\mathbf{a}_t$ will result in reward r_t. However, both agents need corresponding rewards for estimating the action values of $\mathbf{a}_t^c$ and $\mathbf{a}_t^d$ independently. The dilemma is that in the scenario considered, r_t is the result of the combined action of $\mathbf{a}_t^c$, $\mathbf{a}_t^d$ and cannot be split into r_t^c, r_t^d. In other words, from the perspective of the wireless network, the user-side SINR of the STARS network is jointly determined by active beamforming, reflection beamforming, and transmission beamforming. As a consequence, we cannot solely and unambiguously attribute any gain or loss. If r_t is regarded as a common reward for both two agents, the agents will authenticate the MDP transitions as $\{\mathbf{s}_t, \mathbf{a}_t^c, r_t, \mathbf{s}_{t+1}\}$ and $\{\mathbf{s}_t, \mathbf{a}_t^d, r_t, \mathbf{s}_{t+1}\}$. Unfortunately, these transitions are not correct, as in fact the correct transitions have to be formulated as $\{\mathbf{s}_t, \mathbf{a}_t, r_t, \mathbf{s}_{t+1}\}$ or $\{\mathbf{s}_t, \mathbf{a}_t^c, r_t^c, \mathbf{s}_{t+1}^c\}$, $\{\mathbf{s}_t, \mathbf{a}_t^d, r_t^d, \mathbf{s}_{t+1}^d\}$. In order to resolve this dilemma, inspired by Li et al. (2021c), we artificially add an inner environment to formulate the MDPs for a pair of agents.

As shown in Fig. 6.4, an extra environment is derived from the original environment, and the environments refer to the "inner environment" and "outer environment" to distinguish them. Thus, the DDPG agent interacts with the outer environment, and the DQN agent interacts with the inner environment. The outer

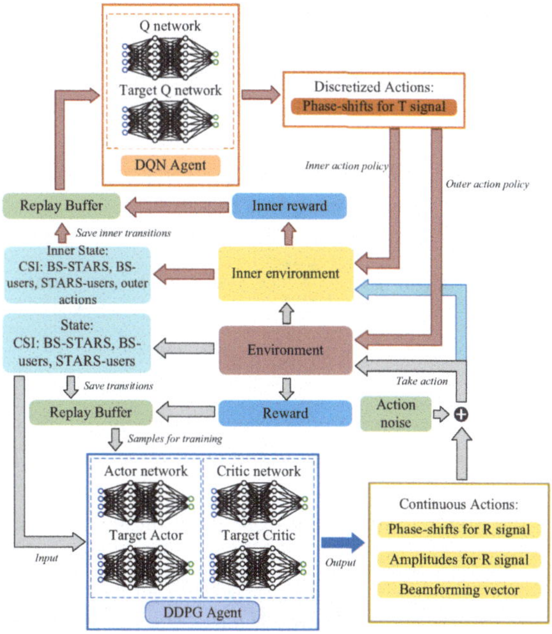

Fig. 6.4 Flow diagram of the proposed DDPG-DQN algorithm

environment represents the STARS-aided wireless network in reality, and the inner environment is only a fictitious environment that contains the knowledge of the outer environment for the agent's training. From the perspective of the DQN agent, the inner environment can be regarded as a collection of the outer environment and the DDPG agent. Based on this framework, for the DQN agent and the inner environment, the MDP can be formulated as $\{\mathbf{s}_{i,t}, \mathbf{a}_{i,t}^d, r_{i,t}, \mathbf{s}_{i,t+1}\}$ and for the outer environment as $\{\mathbf{s}_{o,t}, \mathbf{a}_{o,t}^c, r_{o,t}, \mathbf{s}_{o,t+1}\}$. In the remainder of this section, we discuss the details of the DQN and DDPG agents, respectively.

6.2.3.2 Inner Environment and the DQN Agent

1. *DQN Training*: As a value-based RL algorithm, the DQN algorithm identifies the action values as (6.19), since the DQN agent also aims for the maximum long-term reward. In the inner environment, the Q function is given by

$$Q(\mathbf{s}_{i,t}, \mathbf{a}_{i,t}^d) = \mathbb{E}\big[r(\mathbf{s}_{i,t}, \mathbf{a}_{i,t}^d) + \gamma Q(\mathbf{s}_{i,t+1}, \mathbf{a}_{i,t+1})\big]. \tag{6.35}$$

In the training process, the Q value for the actions has to be updated in each step by following

$$\begin{aligned} Q_{t+1}(\mathbf{s}_{i,t}, \mathbf{a}_{i,t}^d) &\leftarrow (1-\alpha) Q_t(\mathbf{s}_{i,t}, \mathbf{a}_{i,t}^d) \\ &\quad + \alpha[r_{i,t} + \gamma \max Q_t(\mathbf{s}_{i,t+1}, \mathbf{a}_{i,t+1}^d)], \end{aligned} \tag{6.36}$$

where α represents the learning rate ($0 < \alpha \le 1$). In the DQN algorithm, the Q function is approximated by a deep neural network (DNN) having the parameter vector $\boldsymbol{\omega}^Q$, and we have $Q(\mathbf{s}_{i,t}, \mathbf{a}_{i,t}^d) \approx Q(\mathbf{s}_{i,t}, \mathbf{a}_{i,t}^d|\omega^Q)$. In order to accurately fit the Q function, the DNN has to be appreciatively trained. Similar to the DDPG algorithm, the transitions of the DQN agent are stored in the DQN memory buffer, and we can use the memory replay technique for training the DNN by the following loss function:

$$\begin{aligned} L(\omega^Q) = \frac{1}{e}\sum_{e}\big[r_{i,t}(\mathbf{s}_{i,t}, \mathbf{a}_{i,t}^d) + \gamma \max_{\mathbf{a}_{i,t+1}\in\mathbf{A}^d} Q'(\mathbf{s}_{i,t+1}, \mathbf{a}_{i,t+1}|\omega_t^{Q'}) \\ - Q(\mathbf{s}_{i,t}, \mathbf{a}_{i,t}|\omega_t^Q)\big]^2. \end{aligned} \tag{6.37}$$

2. *Inner State and Action Space*: As designed above, the DQN agent regards the DDPG agent as a part of the inner environment. Therefore, the state of the inner environment consists of two parts, including the state inherited from the outer environment and the actions $\mathbf{a}_t^c$ of the DDPG agent. Thus the inner state can be formulated as

$$\mathbf{s}_{i,t} = \{\boldsymbol{H}_{b,r,t}, \boldsymbol{H}_{b,R,t}, \boldsymbol{H}_{b,T,t}, \boldsymbol{H}_{r,R,t}, \boldsymbol{H}_{r,T,t}, \mathbf{a}_{o,t}^c\}. \tag{6.38}$$

Algorithm 6 Joint DDPG-DQN algorithm

```
Initialize the outer environment and the DDPG agent with the actor network ω^μ, critic network
  ω^Q, target actor network ω^{μ'}, target critic network ω^{Q'}.
Initialize the DQN agent with the deep Q network ω^Q and target Q network ω^{Q'}.
for each episode do
    Reinitialize the outer environment to s_{t=0}.
    for each step in t_0 ≤ t ≤ T do
        Observing s_{o,t} and DDPG agent choose a^c_{o,t} according to (6.25).
        DQN agent choose inner action a^d_{i,t} with (6.39).
        Execute action a^d_{i,t} in the inner environment.
        Calculated the reward r_{i,t} and observe the next state s_{i,t+1}.
        Record e{s_{i,t}, a^d_{i,t}, r_{i,t}, s_{i,t+1}}.
        Sample random transitions of e from DQN memory.
        Train ω^Q with (6.37).
        DQN agent choose outer action a^d_{o,t} with (6.40).
        Execute a^c_{o,t} and a^d_{o,t} in the outer environment.
        Calculated the reward r_{o,t} and observe the next state s_{r,t+1}.
        Record e{s_{o,t}, a_{o,t}, r_{o,t}, s_{o,t+1}} in DDPG memory buffer.
        Random sample a batch of transaction e from the memory buffer.
        Calculate target according to (6.22).
        Train critic network Q(s_t, ω_t^Q) with a gradient descent step (6.21).
        Train actor network μ(s_t, ω_t^μ) with (6.23).
        Update the target networks: ω_t^{μ'} ← (1 − τ)ω_t^{μ'} + τω_t^μ,
          ω_t^{Q'} ← (1 − τ)ω_t^{Q'} + τω_t^Q, ω_t^{Q'} ← (1 − τ)ω_t^{Q'} + τω_t^Q.
        s_{o,t} ← s_{o,t+1}
    end for
end for
```

The action space of the DQN agent is $\mathbf{a}^d_{i,t} = \boldsymbol{\Theta}_T$ since the DQN agent is only responsible for the discrete actions. The state space and action indicate a pair of main facts. Primarily, the decision of the DDPG agent has to be known to the DQN agent, when it is interacting with the inner environment. Additionally, it also suggests that the DQN agent has to "interrupt" the interactions of the DDPG agent. Specifically, as shown in Algorithm 6, the timing of activating the inner environment and the DQN agent is after the action selection of the DDPG agent, but before the outer environment's execution of the action, since the transition of the outer environment relies on $\mathbf{a}_{o,t} = \{\mathbf{a}^c_{o,t}, \mathbf{a}^d_{o,t}\}$.

3. *Action Policy*: For conventional DQN agents, the $\epsilon -$greedy action policy constitutes an efficient technique of carefully balancing exploration and exploitation. To elaborate, the $\epsilon -$greedy policy authorizes the DQN agent to choose a random action with the probability of ϵ, and the optimal action with a probability of $1 - \epsilon$. However, it is worth noting that in the joint DDPG-DQN algorithm, the DQN agent has to output a pair of actions one for the inner and one for the outer environment. Action $\mathbf{a}^d_{i,t}$ for the inner environment is adopted to form the inner transitions for the training process of the DQN agent. On the other hand, the action $\mathbf{a}^d_{o,t}$ produced by the DQN for the outer environment has to assist the

DDPG agent. Therefore, once the $\epsilon-$greedy action policy is adopted in the outer environment, it may impose interference on $r_{o,t}$ and lead to an estimation error on the Q value of $\mathbf{a}^c_{o,t}$. Thus, the action policy of the DQN agent can be formulated as

$$\mathbf{a}^d_{i,t} = \begin{cases} \text{random action}, & \epsilon, \\ \arg\max_{\mathbf{a}_{i,t}\in\mathbf{A}^d} Q(\mathbf{s}_{i,t+1}, \mathbf{a}_{i,t}|\boldsymbol{\omega}^Q_t), & 1-\epsilon, \end{cases} \tag{6.39}$$

$$\mathbf{a}^d_{o,t} = \arg\max_{\mathbf{a}_{i,t}\in\mathbf{A}^d} Q(\mathbf{s}_{o,t}, \mathbf{a}_{o,t}|\boldsymbol{\omega}^Q_t). \tag{6.40}$$

Remark 6.2 In the joint DDPG-DQN algorithm, the DQN is trained in the inner environment. The DQN agent produces actions for the outer environment, which may be viewed as a service instead of training for the DQN. Therefore, we choose the ϵ – greedy action policy in the inner environment to train the DQN agent and the optimal action policy in the outer environment to obtain optimal actions.

4. *Reward Function*: Since the reward function depends on the optimization goal and on the constraints, similarly to (6.34), the reward function of the DQN agent is also given by

$$r_{i,t} = \sum_{k=1, R_{k,t}>R_c}^{K} \dot{r} - \sum_{t=1}^{T}\sum_{k=1}^{K} \hat{r} \parallel \mathbf{w}^2_{k,t} \parallel, \tag{6.41}$$

where the constant settings have to be consistent with the corresponding discussion of Sect. 6.2.2.4.

6.2.3.3 Outer Environment and the DDPG Agent

The principle and algorithm's flow of the outer environment and the DDPG agent are the same as in Sect. 6.2.2, and thus we do not repeat the principle and training process here but highlight the difference. For the DDPG agent, state $\mathbf{s}_{o,t}$ is the same as (6.18), and the reward is $r_{o,t} = \sum_{k=1,R_{k,t}>R_c}^{K} \dot{r} - \sum_{t=1}^{T}\sum_{k=1}^{K} \hat{r} \parallel \mathbf{w}^2_k \parallel$, but the difference is that the action space is reduced to $\mathbf{a}^c_{o,t} = \{\mathbf{a}^{\mathbf{w}}_t, \mathbf{a}^{\boldsymbol{\Theta}_R}_t, \mathbf{a}^{\boldsymbol{\beta}}_t\}$. Hence, for any TS t, the optimized value $\{\mathbf{w}_t, \boldsymbol{\Theta}_{R,t}, \boldsymbol{\beta}_t\}$ can be completely covered by $\mathbf{a}^c_{o,t}$ using the mapping approach described in (6.26), (6.27), and (6.28).

In the joint DDPG-DQN algorithm, since both agents are employed and each has its own DNNs, their optimal DNN scales are likely to be smaller than those of the hybrid DDPG algorithm. For the DDPG agent in the joint algorithm, we adopt the same structure as for the DNNs of the hybrid DDPG algorithm, not only as its optimal experimental structure but also to ensure the fairness of the performance comparison.

6.2.3.4 Discussions

Although facing an identical formulated problem, Solution I and Solution II revealed two disparate measures. The hybrid DDPG approach aimed to achieve hybrid control via a carefully tailored mapping function without any additional DNN or other processes. The advantage of this scheme is that its complexity is not increased. By contrast, since it has a lower output dimension than the DDPG agent, i.e., $|\mathbf{a}_n^h| < |\mathbf{a}_n^c| + |\mathbf{a}_n^d|$, it can employ a DNN with fewer trainable parameters, thereby reducing the complexity of the training process. The joint DDPG-DQN algorithm employs a pair of agents having different capabilities for jointly solving the problem having a hybrid action space. Since an extra DQN agent has to be harnessed, given the complexity of the DDPG and the DQN algorithm (Qiu et al., 2019), the total complexity becomes significantly higher than that of the hybrid DDPG scheme.

The complexity of the DDPG algorithm depends on the specification of the employed DNN. Assuming that an actor network having I layers is employed, while each layer contains ω_i^μ nodes, the complexity of propagation is given by $\sum_{i=0}^{I} \omega_i^\mu \omega_{i+1}^\mu$. Given the number of nodes in the actor network as ω_b^μ for BN layers, ω_r^μ for "relu" layers, and ω_t^μ for "tanh" layers, according to Qiu et al. (2019), the required number of floating-point operations is given by $5\omega_\mathrm{b}^\mu + \omega_\mathrm{r}^\mu + 6\omega_\mathrm{t}^\mu$. Applying the same theory for the critic networks, the complexity of a single prediction and training step can be formulated by $O(\sum_{i=0}^{I} \omega_i^\mu \omega_{i+1}^\mu + \sum_{i=0}^{I} \omega_i^Q \omega_{i+1}^Q + 5\omega_\mathrm{b}^\mu + \omega_\mathrm{r}^\mu + 6\omega_\mathrm{t}^\mu + 5\omega_\mathrm{b}^Q + \omega_\mathrm{r}^Q + 6\omega_\mathrm{t}^Q)$. Since we usually have $\omega_i \gg 5$, the computational complexity can be approximated by $O(\sum_{i=0}^{I} \omega_i^\mu \omega_{i+1}^\mu + \sum_{i=0}^{I} \omega_i^Q \omega_{i+1}^Q)$. As mentioned, since the hybrid DDPG algorithm has the smaller output and DNN scale, we have $\omega_i^{\mu,h} < \omega_i^\mu$. Thus, the hybrid DDPG algorithm has lower complexity than the conventional DDPG algorithm.

The complexity of the joint DDPG-DQN algorithm is partially due to the DDPG agent and partially from the DQN agent. The complexity of the DQN agent is given by $O(3\sum_{i=0}^{I} \omega_i^Q \omega_{i+1}^Q)$ as suggested in Li et al. (2021b). The overall complexity of the joint DDPG-DQN algorithm is $O(\sum_{i=0}^{I} \omega_i^\mu \omega_{i+1}^\mu + \sum_{i=0}^{I} \omega_i^Q \omega_{i+1}^Q + 3\sum_{i=0}^{I} \omega_i^Q \omega_{i+1}^Q)$. Therefore, the joint DDPG-DQN algorithm only has an increased complexity compared to the hybrid DDPG algorithm if the DQN agent has a significantly smaller ω_i^Q.

6.2.4 *Numerical Results*

Let us now consider the performance of the proposed approaches. The performance of the STARS is compared to that of the reflecting-only RIS and double-spliced

RIS. The reflecting-only RIS can only serve R users and fails to provide signal enhancements for the T users. As for the other benchmark, the double-spliced RIS is formed by splicing a pair of RISs facing in opposite directions, where the total number of elements is set to N. This also ensures the fairness of comparisons. The double-spliced RIS can also be equivalently regarded as a STARS employing the ES protocol, but the proportion of transmission elements and reflection elements is fixed to 1. In terms of optimization algorithms, the DDPG algorithm having a partial discrete action space is employed as the baseline. The performance of the hybrid DDPG algorithm and DDPG-DQN algorithm is compared.

As for the simulation parameters, the STARS is located 4km away from the BS, where the T users and R users are randomly distributed on both sides of the STARS. The simulated transmission process is 30 seconds, and we assume that the block-fading envelope of the channel varies once per second. For the intelligent agent, we employ the "Adam" optimizer for both the hybrid DDPG and the DDPG-DQN algorithms. The default learning rates for all agents, including the learning rate for the DQN, and the actor/critic learning rates for the DDPG agent are set to be 3×10^{-4}. For the DDPG agent, the actor network has a single activation layer, and the critic network has two activation layers. As for the DQN agent, the deep Q network has 1–2 activation layer(s). The activation function of the hidden layers is "relu" in both algorithms. As mentioned, the size of the activation layers has to be determined by the complexity of the communication system. Empirically, in the majority of realizations, we equip the actor and critic network with hidden layer(s) containing 256–512 neuron nodes, and the DQN agent employs hidden layer(s) with 200–300 neuron nodes. The default parameters used in the simulations are listed in Table 6.1.

Figure 6.5 presents the reward obtained by the proposed algorithms having different learning rates. Based on the rewards exhibited by the different learning

Table 6.1 Default parameters

Parameter	Description	Value
M	The number of antennas	4
N	STARS element number	12
f_c	Carrier frequency	5 GHz
B	Bandwidth	1 MHz
K	The number of users	4
P_{max}	Maximum power per antenna	29 dBm
K_{AR}, K_{RU}	Rician factors	3 dB
σ	Noise power density	−95.2 dBm/MHz
r	Replay buffer size	10,000
γ	Discount factor	1
e_{DDPG}	Batch size	32 samples
e_{DQN}	Batch size	32 samples
τ_{DDPG}	Target update rate	0.002
τ_{DQN}	Target update rate	0.003

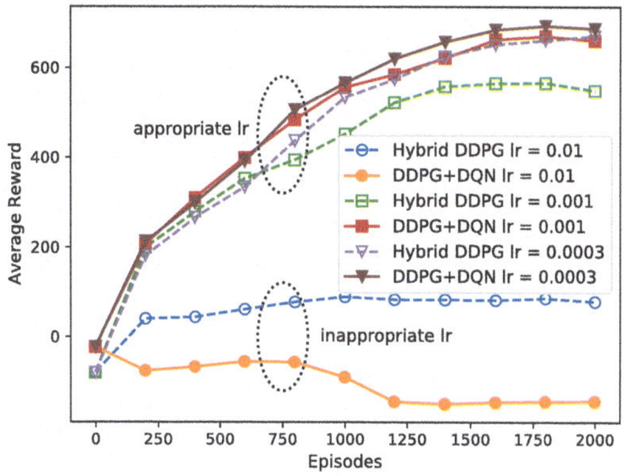

Fig. 6.5 Obtained reward of hybrid DDPG algorithm and DDPG-DQN algorithm with different learning rates

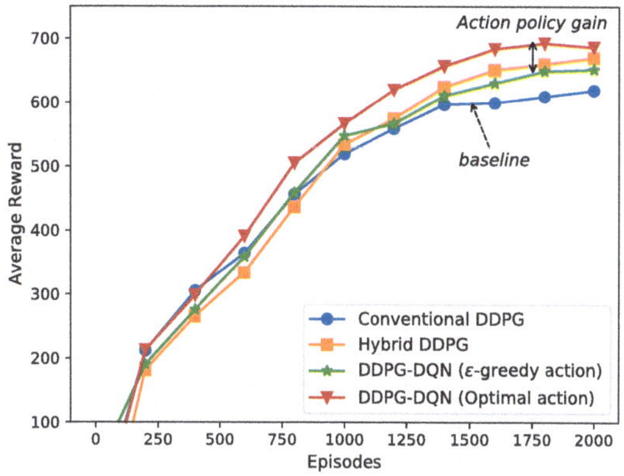

Fig. 6.6 Performances of different algorithms for STARS

rates, we can draw two main conclusions. On the one hand, both the hybrid DDPG and the DDPG-DQN schemes converge if the appropriate learning rate is selected. It has to be clarified that, in general, a higher learning rate leads to faster convergence. However, the curves lr = 0.001 and lr = 0.0003 of Fig. 6.5 exhibit similar convergence rates since we invoked decaying action noise and exploration rate in the (hybrid) DDPG and DQN agents, which affects the reward obtained. On the flip side, the joint DDPG-DQN approach has an excellent capability of obtaining rewards. According to (6.34), the rewards obtained by the agents represent the overall performance of the wireless links, which translate into superior user satisfaction or reduced energy consumption.

The rewards obtained by the different algorithms are plotted in Fig. 6.6. After a training process, the conventional DDPG algorithm used as the baseline has achieved an average reward of about 590, which is inferior to the proposed algorithms. The DDPG-DQN algorithm having optimal action output achieved a slight advantage of about 3–6% compared to that of the hybrid DDPG algorithm, but the price of this performance gain is that the DDQP-DQN algorithm has higher complexity, since two agents are employed. Although the hybrid DDPG scheme is slightly inferior in terms of optimality, it provides a low-complexity rapidly

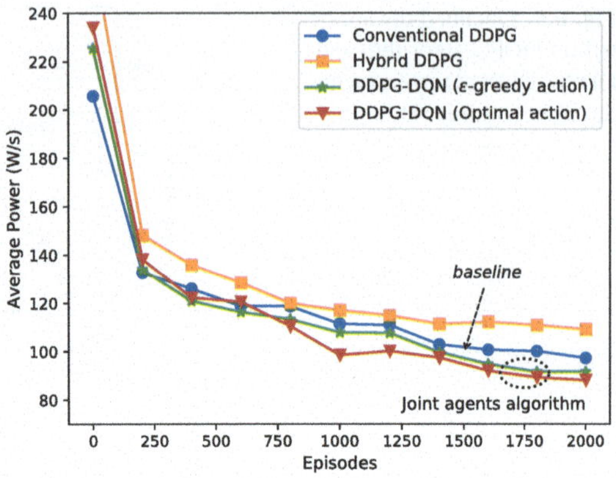

Fig. 6.7 Power consumption of different algorithms for STARS

converging solution. Observing the two branches of the DDPG-DQN scheme, it can be observed that compared to the scheme in Li et al. (2021c), the separated action policy of inner and outer environments has achieved superior performance, which supports the arguments in Remark 6.2.

Figure 6.7 shows the performance of the proposed algorithm from the perspective of power minimization. In the early stage of training, the agent executes fairly random actions, which can be considered as a chaotic system and its transmit power is relatively high. By contrast, for well-trained scenarios, the joint DDPG-DQN scheme has achieved superior power minimization effects regardless of the action policy, which is in line with their trends in the reward analyses. The hybrid DDPG algorithm consumes more power than the DDPG-DQN algorithm, which is why it obtains fewer rewards. What is worth noting is that the baseline algorithm requires lower consumption than the hybrid DDPG algorithm. Recalling reward function (6.34) reveals that the solution provided by the DDPG agent does not satisfy (6.17d) even in some feasible conditions. Therefore, we can speculate that the DDPG algorithm relying on discretized actions can result in completely different actions on both sides of the threshold, which may lead to some action errors and may result in violations of the QoS constraint.

The comparison between STARS, double-spliced RIS, and reflecting-only RIS is presented in Fig. 6.8. STARS has achieved the best overall performance in all training stages, and the double-spliced RIS has about 7% performance disadvantages over the STARS. This phenomenon confirms the conclusion that even if the transmitted and reflected signals of the STARS are constrained by each other, the STARS can provide further signal enhancement for users than double-spliced RISs, since the STARS has higher multipath gain. The reflecting-only RIS has achieved a lower reward than the scheme having double-side coverage since the reflecting-only RIS is not capable of serving T users. In terms of the number of antennas, the simulation results prove that the proposed model and algorithm have general applicability. Furthermore, the multi-antenna gain of the STARS is higher than that of the reflecting-only RIS.

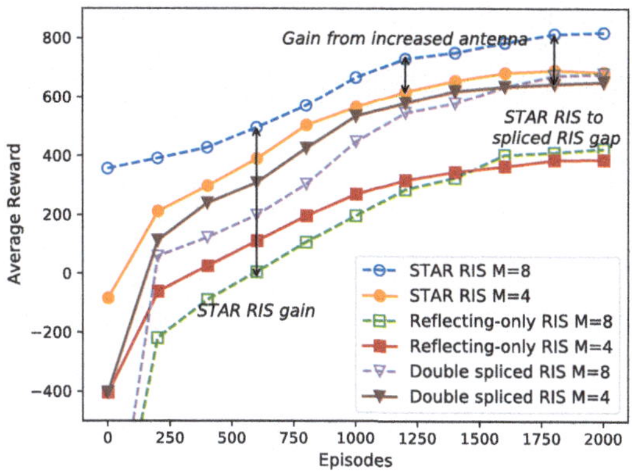

Fig. 6.8 Performance comparison between STARS, reflecting-only RIS, and double-spliced RIS

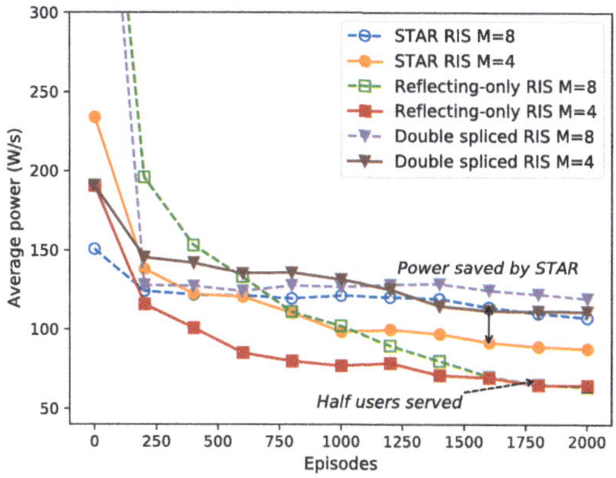

Fig. 6.9 Power consumption of STARS, reflecting-only RIS, and double-spliced RIS

The supporting evidence for Fig. 6.8 is provided by Fig. 6.9, which reveals the transmission power consumption of different types of RISs. The energy consumption characteristics shown by the STARS and double-spliced RIS follow their reward trend. It can be observed that since the STARS can have a higher number of STARS elements to serve users, it obtained higher gains than the double-spliced RIS. Thus, the BS dissipates less transmission power to meet the data rate requirements of users. It is worth paying attention to the fact that the reflecting-only RIS has the lowest power consumption in Fig. 6.9. However, this does not suggest that it has an overall favorable energy efficiency (EE), since it only serves about half the users. Based on similar logic, in the case of $M = 8$, the power consumption seen in Fig. 6.9 is higher than for $M = 4$, since the feasibility of the constraint (6.17d) has changed due to the increased number of antennas.

Figure 6.10 plots the reward against the number of elements. Upon increasing the number of elements, the rewards have also been improved to varying degrees, which is in line with the theoretical expectations of the diversity gain. The result indicates that the performance of the proposed algorithm is not significantly affected by the action dimension (element number). Finally, it can be observed in Fig. 6.10 that the gains obtained by the double-sided coverage of RISs are more significant than those of the single-sided RISs.

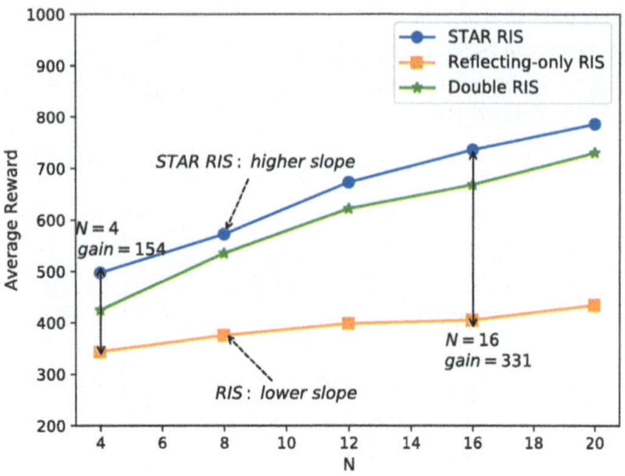

Fig. 6.10 Performances against the number of STARS/reflecting elements

6.3 Distributed Learning Empowered STARS

In this section, we study STARS-aided downlink multicell NOMA networks. For such communication scenarios having a number of terminal devices and BSs, it is desirable to adopt the distributed learning framework to release the resource constraints, lengthy training process, and privacy challenges. To efficiently configure a large number of elements and achieve the desired bidirectional beams for STARS, the DRL algorithms are considered capable of making decisions according to specific scenarios and CSI to achieve long-term intelligent control of the STARS and adaptive resource allocation (Chafaa et al., 2022). However, there are still several challenges to applying distributed DRL for STARS in multicell wireless networks, which are discussed as follows:

- ***Challenge 1:*** Since the STARS elements generally do not have any power amplification functions and the incident signal has experienced fading before arriving at the STARS, the transmitted or reflected signal from every single element has limited energy (Najafi et al., 2021). Hence, in order to ensure sufficient strength of the reconfigured signal, a large-scale STARS with a massive number of elements is necessary. Once an enormous number of elements are employed in the STARS, the optimization complexity of their TRCs will increase significantly, which leads to challenges for invoking the DRL algorithm. The resultant damages include but are not limited to the increase of the action dimension, the extension of the training time, the difficulty in hyperparameter setting, and finally the decline in optimality.
- ***Challenge 2:*** One practical problem is that ML algorithms require high-performance processing equipment and considerable training time (Abels et al., 2019). The distributed deployment scheme is considered to have compatibility with cellular communication networks, which enables agents to contribute their training experience to condense the training load for each agent. Although some studies such as Heess et al. (2017) demonstrated that distributed DRL schemes are feasible, they often require frequent exchange of parameters, resulting in

a huge communication overhead and synchronization problems. Then, with a number of agents and various scales of DNN models, how to manage the exchange of the models in a succinct manner is still a challenge.

- ***Challenge 3:*** Since the DRL algorithm is sensitive to the state dimension, the action dimension, and the scale of the DNN, it suggests that the agent needs to dynamically switch the running models according to the current user status in the cellular. Unfortunately, the number of users in a cellular tends to change frequently, if the agents train a new model based on the current state during each model switching, the system performance will be damaged and even outage due to the untrained model. Therefore, how to warm start or switch the model to avoid performance degradation becomes a challenge.

In the following, we will develop a distributed DRL approach to empower the intelligent STARS-aided downlink multicell NOMA networks as a candidate to provide accessibility and coverage for the future wireless network.

6.3.1 System Model and Problem Formulation

As shown in Fig. 6.11, we consider STARS-aided downlink multicell NOMA networks, where each cellular comprises a BS $b, b \leq B$, K_b users, and a STARS $r, r \leq R$ equipped with N STARS elements as illustrated in Fig. 6.11. The STARS is equipped with a plenty of STARS elements, which is consequent in a challenge for the TRC control (Yu et al., 2021). We denote the locations of the BS, RIS, and users as $(x_b, y_b, z_b)^T$, $(x_r, y_r, z_r)^T$, and $\left(x_{k_b}, y_{k_b}, z_{k_b}\right)^T$, respectively. We assume that the STARS is capable of achieving 360° coverage by the transmitted and reflected signal to serve users around the STARS. Each user k_b receives transmitted or reflected signal depending on its location, where users having $x_{k_b} < x_r$ are served by the reflected signal and users having $x_{k_b} > x_r$ are covered by the transmitted signal.

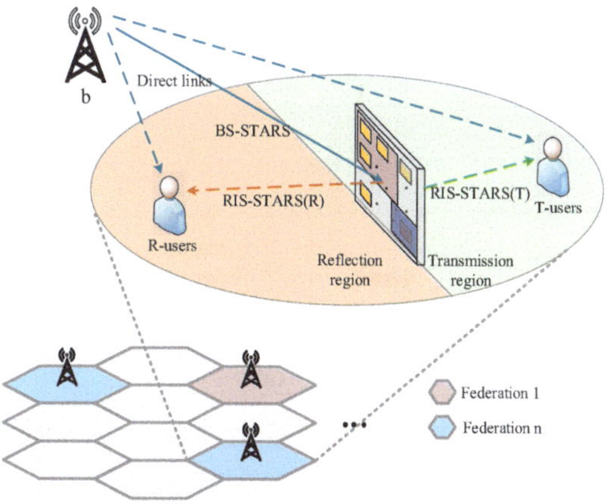

Fig. 6.11 System model of STARS-aided downlink multicell NOMA networks

Without loss of generality, we assume different BSs may be associated with a different number of users K_b, but STARSs have the same value of N and M as the communication equipment is standardized. In each cell, the intelligent agent is deployed on a controller, which is connected with both the BS and the STARS by wired links to achieve joint control. Since BSs and controllers have wired links, agents can transfer commands and parameters without spectrum overhead. Multiple BSs in the considered scenario do not necessarily have similar or adjacent locations. The BSs are observed and managed according to the number of active users they have, and thus, the cells considered can be located in different locations, cities, or even countries. Hence, due to the separate spatial location, we need to clarify that the scope of this chapter does not include suppression of intercell interference between adjacent cells and average intercell interference is considered among the cellular instead.

6.3.1.1 Proposed Tile-Based STARS Operation

Here, we consider the ES and MS operating protocols for STARSs. In order to efficiently resolve the passive beamforming problem for the STARS, we propose a tile-based STARS beamforming scheme, which partitions the STARS elements into M subsets (tiles), and elements in the same tile have a common TRC. We will briefly introduce the ES protocol and the MS protocol and discuss their corresponding tile partition schemes, respectively.

1. *ES STARS Partitioning Scheme*: As shown in Fig. 6.12a, since each STARS element in ES simultaneously provides transmitted and reflected signals, the TRCs of each STARS element can be denoted by $\beta_{T,n}e^{j\theta_{T,n}}$ and $\beta_{R,n}e^{j\theta_{R,n}}$, respectively. Here, we still consider the practical coupled phase-shift model, where the coupling between the TRCs for a given ES STARS element n is given by

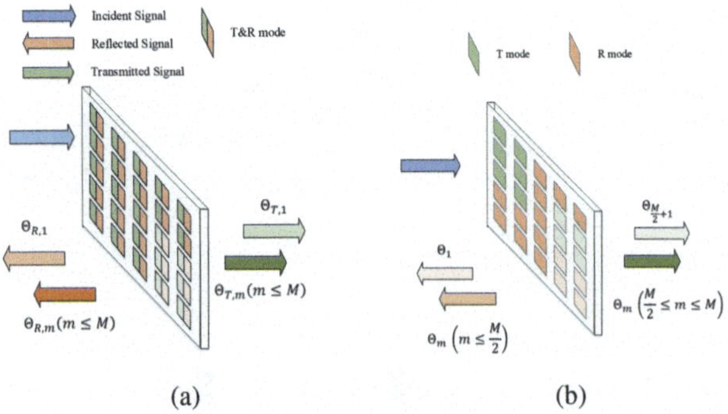

Fig. 6.12 Tile-based STARS. (**a**) Tile-based ES STARS. (**b**) Tile-based MS STARS

$$\beta_{T,n}\beta_{R,n}\cos(\theta_{R,n} - \theta_{T,n}) = 0, \tag{6.42}$$

which suggests $\Delta\theta_n = \theta_{R,n} - \theta_{T,n}$, $\Delta\theta_n \in \{-\pi/2, \pi/2\}$. Meanwhile, according to the conservation of energy, we have the relationship between $\beta_{T,n}$and $\beta_{R,n}$ that follows $\beta_{R,n} = \sqrt{1-\beta_{T,n}^2}$. Therefore, for the ES STARS model, the TRCs matrices are given by

$$\boldsymbol{\Theta}_T^{\text{ES}} = \text{diag}\left(\beta_{T,1}e^{j\theta_{T,1}}, \beta_{T,2}e^{j\theta_{T,2}}, \cdots, \beta_{T,N}e^{j\theta_{T,N}}\right), \tag{6.43}$$

$$\boldsymbol{\Theta}_R^{\text{ES}} = \text{diag}\left(\sqrt{1-\beta_{T,1}^2}e^{j\theta_{T,1}+\Delta\theta_1}, \cdots, \sqrt{1-\beta_{T,N}^2}e^{j\theta_{T,N}+\Delta\theta_N}\right). \tag{6.44}$$

For the tile-based ES STARS beamforming scheme, we have to assign all N elements into M tiles. We define the element partition matrix $\boldsymbol{\Lambda}_m^{\text{ES}} \in \mathbb{C}^{N\times N}, m \leq M$ for the m-th tile. $\boldsymbol{\Lambda}_m^{\text{ES}}$ is a diagonal matrix, while $\boldsymbol{\Lambda}_m^{\text{ES}}[n,n] = 1$ indicates the element n is in the tile m, and $\boldsymbol{\Lambda}_m^{\text{ES}}[n,n] = 0$ represents the element n is not in the tile m. For each tile, it has the same TRC, $\tilde{\boldsymbol{\Theta}}_{T,m} = \tilde{\beta}_{T,m}e^{j\tilde{\theta}_{T,m}}$ and $\tilde{\boldsymbol{\Theta}}_{R,m} = \sqrt{1-\tilde{\beta}_{T,m}^2}e^{j\tilde{\theta}_{T,m}+j\Delta\tilde{\theta}_m}$. The entire TRC matrices in the tile-based ES STARS are given by

$$\tilde{\boldsymbol{\Theta}}_T^{\text{ES}} = \sum_{m=1}^{M} \tilde{\boldsymbol{\Theta}}_{T,m}\boldsymbol{\Lambda}_m^{\text{ES}}, \tag{6.45}$$

$$\tilde{\boldsymbol{\Theta}}_R^{\text{ES}} = \sum_{m=1}^{M} \tilde{\boldsymbol{\Theta}}_{R,m}\boldsymbol{\Lambda}_m^{\text{ES}}. \tag{6.46}$$

2. *MS STARS Partitioning Scheme*: In the MS protocol, the STARS consists of two groups of elements, where one group operates in the reflection function and the other one is responsible for the transmissions. Each MS element employed in the STARS can be either working in reflection mode or transmission mode as shown in Fig. 6.12b, and the resulting "on-off" working mode makes MS protocol easy to implement in practice. Assuming there are N_T elements working in the transmission mode and N_R elements in the reflection mode, we have $N_T + N_R = N$. Then, the TRC matrices can be given by

$$\boldsymbol{\Theta}_T^{\text{MS}} = \text{diag}\left(\beta_{T,1}e^{j\theta_{T,1}}, \beta_{T,2}e^{j\theta_{T,2}}, \cdots, \beta_{T,n}e^{j\theta_{T,N}}\right), \tag{6.47}$$

$$\boldsymbol{\Theta}_R^{\text{MS}} = \text{diag}\left(\beta_{R,1}e^{j\theta_{R,1}}, \beta_{R,2}e^{j\theta_{R,2}}, \cdots, \beta_{R,N}e^{j\theta_{R,N}}\right). \tag{6.48}$$

For the MS STARS, there is no additional couple phase-shift constraint, i.e., $\theta_{T,n}, \theta_{T,n} \in [0, 2\pi)$.

For the tile-based MS STARS, we assume to assign N_T elements into $M/2$ tiles for transmission and N_R elements into another $M/2$ tiles for reflection. Please note that the number of transmission tiles and reflection tiles can be set arbitrarily according to the need. Here, we set them to $M/2$ without loss of generality. We define the element partition matrix $\mathbf{\Lambda}_{T,m}^{\text{MS}} \in \mathbb{C}^{N\times N}$, $m \le M/2$ and $\mathbf{\Lambda}_{R,m}^{\text{MS}} \in \mathbb{C}^{N\times N}$, $M/2 < m \le M$ for the transmission and reflection, respectively. Similarly, $\mathbf{\Lambda}_{T,m}^{\text{MS}}, \mathbf{\Lambda}_{R,m}^{\text{MS}}$ are diagonal matrices, while $\mathbf{\Lambda}_{T(R),m}^{\text{MS}}[n,n] = 1$ indicates element n is in tile m, and $\mathbf{\Lambda}_{T(R),m}^{\text{MS}}[n,n] = 0$ represents element n is not in tile m. In each tile, the elements persist in the same TRC $\tilde{\mathbf{\Theta}}_{T,m}^{\text{MS}}$ or $\tilde{\mathbf{\Theta}}_{R,m}^{\text{MS}}$, and the overall TRC matrices are given by

$$\tilde{\mathbf{\Theta}}_{T}^{\text{MS}} = \sum_{m=1}^{M/2} e^{j\tilde{\theta}_{T,m}} \mathbf{\Lambda}_{T,m}^{\text{MS}}, \tag{6.49}$$

$$\tilde{\mathbf{\Theta}}_{R}^{\text{MS}} = \sum_{m=M/2+1}^{M} e^{j\tilde{\theta}_{R,m}} \mathbf{\Lambda}_{R,m}^{\text{MS}}. \tag{6.50}$$

As MS STARS can be regarded as one special case of ES STARS, in the following, we focus our attention on the tile-based STARS beamforming design for the ES case.

6.3.1.2 Communication Model

The quasi-static block-fading channel is invoked to characterize the channels in the STARS scenario, where the fading blocks are assumed to be a finite set. The channel spanning BS b to the STARS r, which is denoted by $\boldsymbol{h}_{b,r} \in \mathbb{C}^{1\times N}$, is assumed to follow Rician fading since the LoS path is likely to be obtained. The direct channel between the BS and user k is denoted by $h_{b,k}$.[3] The channel linking the STARS to user k is denoted by $\boldsymbol{h}_{r,k} \in \mathbb{C}^{N\times 1}$. Due to the random movement of the users, the LoS path is not guaranteed at the user equipment. Therefore, we assume the channel $h_{b,k}, \boldsymbol{h}_{r,k}$ obeying the Rayleigh fading model. Meanwhile, the distance-based path loss for all channels follows the classic model that $\mathcal{L}(d) = \rho_0 \left(\frac{d}{d_0}\right)^{-\alpha}$ as suggested in Zuo et al. (2022), where ρ_0 represents the path loss at the reference distance d_0, d is the distance between the transceivers, and α is the path loss exponent.

[3] Please note that since the user k is assumed associated with BS b, for concision, the superscript b for user k_b is omitted.

Given the aforementioned models, the BS-user channel and the RIS-user channel are given by $h_{b,k} = \sqrt{\mathcal{L}_{b,k}(d_{b,k})}g_{b,k}$ and $\boldsymbol{h}_{r,k} = \sqrt{\mathcal{L}_{r,k}(d_{r,k})}\boldsymbol{g}_{r,k}$, where $g_{b,k}, \boldsymbol{g}_{r,k} \sim \mathcal{CN}(0, 1)$. Moreover, the channel $\boldsymbol{h}_{b,r}$ can be expressed as

$$\boldsymbol{h}_{b,r} = \sqrt{\mathcal{L}_{b,r}\left(d_{b,r}\right)}\left(\sqrt{\frac{K}{K+1}}\boldsymbol{l}_{b,r} + \sqrt{\frac{1}{K+1}}\boldsymbol{g}_{b,r}\right), \tag{6.51}$$

where K denotes the Rician factor, $\mathbf{g}_{b,r} \sim \mathcal{CN}(0, 1)$ denotes the scattered path component of the signal, and $\boldsymbol{l}_{b,r}$ is the LoS path component. Since two elements in the same tile may have significant spatial distance, in the case that the user is not far enough from the STARS, the AoA and AoD of the signal have to be calculated individually for each element. Explicitly, elements in the tile having the same TRCs will cause errors in the phase shift of elements, and this is one of the costs of the tile-based control. To ensure accuracy, we introduce an accurate AOA/AOD model to evaluate its performance loss. We have $\boldsymbol{l}_{b,r} = \boldsymbol{l}_r(\varphi^A, \psi^A)l_b(\varphi^D)^H, \boldsymbol{l}_r \in \mathbb{C}^{N\times 1}$, where $\varphi^{A/D} \in [0, 2\pi)$ and $\psi^A \in [-2/\pi, 2/\pi)$ represent the azimuth and elevation AOA/AOD, $\varphi^A = \arcsin(\frac{y_r - y_b}{\sqrt{(x_r-x_b)^2+(y_r-y_b)^2}})$, $\varphi^D = \pi/2 - \varphi^A$, and $\psi^A = \arcsin(\frac{z_r - z_b}{\sqrt{(x_r-x_b)^2+(y_r-y_b)^2}})$.

We assume that the STARS has a UPA of elements, where the antenna space and the element space are d_a and d_e. According to Zhang and Zhang (2020), we have

$$l_b = e^{j2\pi d_a \sin(\varphi^D)/\lambda}, \tag{6.52}$$

$$l_r[n] = e^{j2\pi(n-1)d_e(\lfloor n/N_x \rfloor \sin(\varphi^A)\sin(\psi^A)+(n-\lfloor n/N_x \rfloor N_x)\sin(\varphi^A)\cos(\psi^A))/\lambda}, \tag{6.53}$$

where N_x denotes the number of reflective elements in each row and $\lfloor\rfloor$ is the floor function.

Remark 6.3 Since two distant elements are likely to receive signals with different AOAs and require different AODs, it is not wise to assign these elements to the same tile. Therefore, the size of a tile and its element allocation need to take this factor into account.

Without loss of generality, we assume the T users and R users are in one NOMA cluster. According to the principle of the superposition coding (SC), the transmitted signal at the BS b can be given by

$$x_b = \sum_{k=1}^{K} \sqrt{P_{b,k}}x_{b,k}, \tag{6.54}$$

where P_k represents the allocated power for user k. Please note that all the signals are time-varying during the transmission period, and we omit the superscript t for a concise presentation. Since we assumed the users can receive the signal from both the direct link and the cascaded STARS link, the composite channel for user k can be given by

$$\hat{h}_{b,k} = \begin{cases} h_{b,k} + \boldsymbol{h}_{r,k}\tilde{\boldsymbol{\Theta}}_{T,t}\boldsymbol{h}_{b,r}, \forall k \in \mathbb{K}_T, \\ h_{b,k} + \boldsymbol{h}_{r,k}\tilde{\boldsymbol{\Theta}}_{R,t}\boldsymbol{h}_{b,r}, \forall k \in \mathbb{K}_R, \end{cases} \tag{6.55}$$

where $\mathbb{K}_T$ and $\mathbb{K}_R$ represent the set of T and R users, respectively. According to the principle of the SIC, NOMA users in the same cluster can remove a part of the intracluster interference. To ensure a successful SIC for each user, the decoding order has to be correctly determined. As suggested in Yang et al. (2017), the decoding order has to be determined by the channel gain order. Assuming there are two users k, j in the same NOMA cluster, the decoding order is given by $\Omega_j > \Omega_k, \forall |h_j|^2 > |h_k|^2$. Then, the received signal at the user k is given by

$$y_{b,k} = \sqrt{P_{b,k}}\hat{h}_{b,k}x_{b,k} + \sum_{j,\Omega_j > \Omega_k,} \sqrt{P_{b,j}}\hat{h}_{b,j}x_{b,j} + n_0, \tag{6.56}$$

where the first term is the desired signal, the second term is intracluster interference, and n_0 represents the sum of the average intercluster interference and AWGN.

Given the received signal, the SINR of user k can be calculated by

$$\gamma_{b,k} = \frac{P_{b,k} \mid \hat{h}_{b,k} \mid^2}{\sum\limits_{j,\Omega_j > \Omega_k} P_{b,j} \mid \hat{h}_{b,k} \mid^2 + \sigma^2}, \tag{6.57}$$

where σ represents the intercluster interference and noise power. Therefore, the data rate of each user is given by

$$R_{b,k} = W \log_2\left(1 + \gamma_{b,k}\right), \tag{6.58}$$

where W represents the frequency bandwidth.

6.3.1.3 Problem Formulation

We aim for the average throughput maximization of the BSs by jointly optimizing the power allocation P_k and the corresponding tile-based beamforming design $\tilde{\boldsymbol{\Theta}}_{T,m}, \tilde{\boldsymbol{\Theta}}_{R,m}$. The optimization problem can be formulated as

$$\max_{\boldsymbol{\Lambda}_m, \tilde{\boldsymbol{\Theta}}_{T,m}, \tilde{\boldsymbol{\Theta}}_{R,m}, P_{b,k}} \sum_{b=1}^{B}\sum_{k=1}^{K}\sum_{t=1}^{T} R_{b,k}, \tag{6.59a}$$

$$\text{s.t.} \quad 0 \leq \tilde{\theta}_{T,m} < 2\pi, \forall m, \tag{6.59b}$$

$$0 \leq \tilde{\theta}_{R,m} < 2\pi, \forall m, \tag{6.59c}$$

$$\tilde{\beta}_{T,m}\tilde{\beta}_{R,m}\cos(\tilde{\theta}_{R,m} - \tilde{\theta}_{T,m}) = 0, \tag{6.59d}$$

$$\tilde{\beta}_{T,m}^2 + \tilde{\beta}_{R,m}^2 = 1, \forall m, \tag{6.59e}$$

$$R_{b,k} \geq R_{\text{QoS}}, \forall k, \tag{6.59f}$$

$$\sum_{m=1}^{M} \mathbf{\Lambda}_m = \mathbf{I}, \tag{6.59g}$$

$$\sum_{k=1}^{K} P_{b,k} \leq P_{\max}, \tag{6.59h}$$

where constraints (6.59b)–(6.59e) represent the phase-shift and amplitude range for STARS. Constraint (6.59f) is a quality-of-service (QoS) constraint that stipulates the minimum data rate for users. Constraint (6.59g) indicates that each element has to be and only be assigned to a tile. Finally, (6.59h) is the maximum power constraint for the BS. Even if we temporarily neglect the complexity contributed by the optimization of the power allocation for NOMA users and the TRCs of each tile, the choice of $\mathbf{\Lambda}_m$ is an integer choice problem for each. If each element selects its tile in turn, it will result in M^N choices. Considering the massive number of elements in the STARS, that will be a complex and challenging problem. Furthermore, since multiple BSs, random moving users, and a period of transmission time are considered in the scenario, the optimization problem is a dynamic long-term optimization problem. Therefore, we propose a DRL-based distributed ML scheme to solve the formulated problem.

6.3.2 *Distributed Learning Algorithm for STARS*

In the following, we describe the proposed algorithm in detail to solve the formulated optimization problem, where a proximal policy optimization (PPO) algorithm is invoked for the STARS partitioning and corresponding beamforming, and an access-free federated learning (AFFL) multiagent distributed learning framework is proposed for the experience exchange among agents. We will first briefly introduce the principles of the PPO algorithm and describe how to deploy the PPO agent in a single cell, including the design of states, actions, and reward functions. Then, we will expand the scope to the multiagent case. We will explain why the AFFL framework is desired by communication systems, and we will illustrate the principle of the AFFL framework.

6.3.2.1 PPO-Based STARS Partition and Beamforming

The PPO algorithm (Schulman et al., 2017) is invoked to provide solutions for the STARS in each cellular. After the advent of the PPO algorithm, it achieved impressive results in a variety of training environments, becoming one of the current mainstream DRL algorithms (Naughton et al., 2021). As a novel PG algorithm, the PPO algorithm overcomes a common challenge in PG algorithms that they are sensitive to the step size of the policy update, and the oversized update can result in devastating performance degradation. By employing the clipped surrogate objective or the KL-penalized objective, the PPO algorithm successfully constrains the size of a policy update, which ensures its stable convergence and leads to a lower complexity compared to other conventional PG algorithms. Due to the aforementioned superiority, it is employed for the considered optimization problem.

Following the basic principle of the DRL algorithm, the PPO agent has to be trained by interacting with the communication environment. During the transmission period $0 \leq t \leq T$, the agent has to recognize the current $\mathbf{s}_t \in \mathbf{S}$ and choose an action $\mathbf{a}_t \in \mathbf{A}$ according to the policy $\pi(\mathbf{a}_t|\mathbf{s}_t)$. As a consequence of taking action $\mathbf{a}_t$, the agent receives a reward that indicates the quality of action $\mathbf{a}_t$, and the state will transfer to $\mathbf{s}_{t+1} \in \mathbf{S}$. Thus, a crucial factor in successfully using the PPO algorithm to resolve the STARS optimization problem is to determine an appropriate state space $\mathbf{S}$, action space $\mathbf{A}$, and reward function to determine r_t.

1. *State Space*: State space has to contain the relevant information required for STARS partitioning, TRC optimization, and power allocation. The PPO agent has the identical state space whether ES or MS are considered, where the CSI at fading block t is included in the state $\mathbf{s}_t$ as given in (6.60).

$$\mathbf{s}_t = c_0\{\boldsymbol{h}_{b,r,t}, \boldsymbol{h}_{r,k}, h_{b,k,t}\}. \tag{6.60}$$

   In addition, since the path loss in the wireless network can be severe, the value of the elements in $\mathbf{s}_t$ is sometimes close to 0. It can cause the actor and critic network to receive a state vector with insignificant differences or even invalid data types, which is not conducive to the DNN fitting. Hence, empirically an adaptive scaling coefficient c_0 or a normalization needs to be considered in contracting $\mathbf{s}_t$.
2. *Action Space*: The PPO algorithm is capable of handling both discrete and continuous action spaces, and we invoke PPO implementations with continuous action spaces in this chapter. Although the partitioning problem of tiles prefers discrete actions as the number of STARS elements has to be an integer, we adopt a continuous action space for obtaining accurate TRCs and power distribution. Once discrete actions are required, decision thresholds can be incorporated to discretize the sampled continuous actions succinctly. For the conventional element-based RIS beamforming problem, a straightforward action space design as shown in (6.61) is widely acknowledged

$$\mathbf{a}_t = \{\mathbf{a}_t^{\boldsymbol{\Theta}}, \mathbf{a}_t^{p}\}, \tag{6.61}$$

   where $\mathbf{a}_t^{\boldsymbol{\Theta}}$ and $\mathbf{a}_t^{p}$ represent the output action of the agent for the phase shift and the power allocation, respectively. The RC $\boldsymbol{\Theta}$ and the power P_k for each user can be obtained by applying simple linear mapping functions. However, this method leads to an enormous demission of the action space.

For the partitioning-based approach, the element partition matrix $\mathbf{\Lambda}^{T,m}$, $\mathbf{\Lambda}^{R,m}$ cannot be directly given by the agent. Indeed, the agent can assign each element to the tile that it belongs to. Taking the MS protocol as an example, the action space can be designed as

$$\mathbf{a}_t = \{\mathbf{a}_t^{\mathbf{\Lambda}_{T,m}^M}, \mathbf{a}_t^{\mathbf{\Lambda}_{R,m}^M}, \mathbf{a}_t^{\tilde{\mathbf{\Theta}}_T^M}, \mathbf{a}_t^{\tilde{\mathbf{\Theta}}_R^M}, \mathbf{a}_t^p\}, \tag{6.62}$$

where $\mathbf{a}_t^{\mathbf{\Lambda}_{T,m}^M}$, $\mathbf{a}_t^{\mathbf{\Lambda}_{R,m}^M}$ represent the element partition matrices and $\mathbf{a}_t^{\tilde{\mathbf{\Theta}}_T^M}$, $\mathbf{a}_t^{\tilde{\mathbf{\Theta}}_R^M}$ include the information of the TRCs for each tile. Unfortunately, this scheme still has a large action dimension since $\mathbf{a}_t^{\mathbf{\Lambda}_{T,m}^M}$, $\mathbf{a}_t^{\mathbf{\Lambda}_{R,m}^M}$ have a dimension N in total, which is contrary to our original intention of reducing complexity and action space dimension. Moreover, since the adjacent elements $[n, n]$ and $[n+1, n+1]$ in the diagonal matrixes $\mathbf{\Lambda}^{M_R}$, $\mathbf{\Lambda}^{M_T}$ are not guaranteed to be adjacent in the 2D STARS panel, there is a possibility that multiple scattered elements are assigned to one tile. Recalling the discussion in Remark 6.3, such an allocation is not in line with the principles of the propagation.

In order to achieve a low-complexity partition for the STARS, we propose an indirect method to determine the tile. We do not directly seek the attribution of each element but assign the agent to output the horizontal and vertical dividing boundary of the tile. For example, if a STARS plane needs to be divided into $M = M_x \times M_y$ tiles, where M_x and M_y represent the number of horizontal and vertical tiles, respectively, the agents output the $N_{m,x}$ and $N_{m,y}$. Thus we have $\mathbf{\Lambda}^m[n_x, n_y] = 1, N_{m,x} \le n_x \le N_{m+1,x}, N_{m,y} \le n_y \le N_{m+1,y}$. According to the above method, for the ES protocol, the action spaces can be designed as

$$\mathbf{a}_t = \{\mathbf{a}_t^{\mathbf{n}_x^M}, \mathbf{a}_t^{\mathbf{n}_y^M}, \mathbf{a}_t^{\tilde{\mathbf{\Theta}}_R^M}, \mathbf{a}_t^{\Delta\tilde{\mathbf{\Theta}}^M}, \mathbf{a}_t^p\}, \tag{6.63}$$

where $\mathbf{a}_t^{\mathbf{n}_x^M}$, $\mathbf{a}_t^{\mathbf{n}_y^M}$ represent the vectors containing the horizontal and vertical boundary for tiles, and $\mathbf{a}_t^{\tilde{\mathbf{\Theta}}_T^M}$ represents the RC for tiles, and $\mathbf{a}_t^{\Delta\tilde{\mathbf{\Theta}}^M}$ contains the phase-shift difference between the TC and the RC. After harvesting the above actions, we can calculate $\mathbf{\Lambda}^M$, $\tilde{\mathbf{\Theta}}_R^M$, and $\tilde{\mathbf{\Theta}}_T^M$. Please note that for the ES protocol, $\mathbf{\Lambda}_R^M = \mathbf{\Lambda}_T^M = \mathbf{\Lambda}^M$. Then, following (6.45), (6.46), and (6.63), we can calculate TRC for all elements.

Following the similar partitioning logic, the action space of the MS protocol can be given by

$$\mathbf{a}_t = \{\mathbf{a}_t^{\mathbf{n}_x^M}, \mathbf{a}_t^{\mathbf{n}_y^M}, \mathbf{a}_t^{\tilde{\mathbf{\Theta}}_R^M}, \mathbf{a}_t^{\tilde{\mathbf{\Theta}}_T^M}, \mathbf{a}_t^p\}. \tag{6.64}$$

Please note that due to the different features between the ES and MS, in MS protocol we have to distinguish the $\mathbf{\Lambda}_T^M$, $\mathbf{\Lambda}_R^M$, and it needs independent outputs for the $\tilde{\mathbf{\Theta}}_T^M$ and $\tilde{\mathbf{\Theta}}_R^M$.

Remark 6.4 The action dimension of element-based beamforming is $3N + K$, but only about $4M + K$ in tile-based beamforming. Considering the practical case that $N > 100 >> M$, the action dimension of element-based beamforming will be huge, which requires large-scale DNNs and is likely to lead to increasing complexity, slow convergence, and inferior optimality.

3. *Reward Function*: The reward function determines the optimization orientation for the intelligent agent, which will have a major impact on the optimization effect. Therefore, the reward function has to be consistent with the objective function. We define the reward function as

$$r_t = \begin{cases} R_t, & \text{QoS requirement satisfied,} \\ \frac{R_t}{c_p}, & \text{QoS requirement not satisfied,} \end{cases} \tag{6.65}$$

where c_p is a penalty for violation of QoS requirements to enforce the agent to meet the constraint (6.17d).

4. *Training Process*: PPO algorithm genesis forms policy gradient methods, which aims to train stochastic policy $\pi_\omega(\boldsymbol{a}_t|\boldsymbol{s}_t)$ to obtain the optimal parametric probability distribution for actions against $\boldsymbol{s}_t$. A conventional gradient estimator is given by

$$L_t^{\text{PG}}(\omega) = \mathbb{E}\left[\log \pi_\omega(\boldsymbol{a}_t|\boldsymbol{s}_t)\mathcal{A}_t^s\right], \tag{6.66}$$

where $\mathcal{A}_t^s$ is an advantage function at state s_t. Then, for a given length of time frame $0 < t < T$, it can be expressed as

$$\mathcal{A}_t^s = \delta_t + (\gamma\lambda)\delta_{t+1} + \cdots + (\gamma\lambda)^{T-t+1}\delta_{T-1}, \tag{6.67}$$

$$\delta_t^s = r_t + \gamma V_\omega(\boldsymbol{s}_{t+1}) - V_\omega(\boldsymbol{s}_t), \tag{6.68}$$

where γ is the discount factor, and $V_\omega(s)$ represents the state-value function (Schaul et al., 2015). In order to solve the defect that the transition samples of the policy gradient algorithm cannot be reutilized for training, different from the conventional policy gradient algorithm, the PPO algorithm adopts a surrogate objective, which enables the agent to train the current policy π by using samples generated by the previous policy π_{old}, where the surrogate objective is given by

$$L^{\text{SO}} = \mathbb{E}\left[u_t(\omega)\mathcal{A}_t^s\right], \tag{6.69}$$

where

$$u_t(\omega) = \frac{\pi_\theta(\boldsymbol{a}_t|\boldsymbol{s}_t)}{\pi_{\theta_{\text{old}}}(\boldsymbol{a}_t|\boldsymbol{s}_t)}. \tag{6.70}$$

However, maximizing L^{SO} without any constraint is likely to lead to an excessively large update on policy. Therefore, following the design in Schulman et al. (2017), we adopt a loss function that combines the policy surrogate and a value function error term, which is given by

$$L_t(\omega) = \mathbb{E}\left[L_t^{\text{CLIP}}(\omega) - c_1 L_t^{\text{VF}}(\omega) + c_2 S[\pi_\theta](\boldsymbol{s}_t)\right], \tag{6.71}$$

where $c1$ and $c2$ are coefficients, L_t^{VF} is the squared value function error term that

$$L_t^{\text{VF}}(\omega) = \mathbb{E}_t\left[\left(V_\omega(\boldsymbol{s}_t) - V_{\pi_{\omega_{\text{old}}}}(\boldsymbol{s}_t)\right)^2\right], \tag{6.72}$$

and $c_2 S[\pi_\theta](\boldsymbol{s}_t)$ an entropy bonus for exploration. The most distinct contribution of the PPO algorithm is embodied in the first term, which is a clipped surrogate objective proposed in the PPO2 algorithm. It can be given by

$$L_t^{\text{CLIP}}(\omega) = \mathbb{E}\left[\min\left(u_t^s(\omega)\mathcal{A}_t^s, \text{clip}\left(u_t^s(\omega), 1-\epsilon, 1+\epsilon\right)\mathcal{A}_t^s\right)\right], \tag{6.73}$$

where ϵ is a hyperparameter to limit the update on the policy.

With the calculated loss L_t, we can update the DNNs following

$$\omega \leftarrow \omega + \alpha \nabla_\omega L_t(\omega), \tag{6.74}$$

where α represents the learning rate.

6.3.2.2 AFFL Model

The AFFL model invokes both federated learning (FL) and transfer learning (TL), establishing multiple federations according to the network status of each BS and authorizing the agent to freely shuttle between each federation, thereby exempting the time overhead required by retraining the models. As reviewed in the state of the art, several distributed frameworks were proposed for the PPO or other DRL algorithms. These successful implementations proved that incorporating DRL agents into distributed frameworks can effectively reduce training time, distribute computing load to hardware devices, and lay a solid theoretical foundation for distributed RL. However, these schemes may encounter practical difficulties when using them to arm communication systems. Since in a communication system, thousands of BSs having the same technical specifications work in parallel, this feature becomes the technical prerequisite for applying distributed learning among BSs. However, the existing distributed frameworks are not specifically designed for the communication network, which would meet practical issues once they are employed. Taking the distributed proximal policy optimization (DPPO) (Heess et al., 2017) as an example, if the DPPO paradigm is employed to coordinate BSs, then each BS needs to play the role of worker, while the central server will play the role of the chief. Then in each training slot, the chief needs to wait until it receives the policy gradient from all workers before starting training. This scheme not only requires frequent data exchanges between BSs and the central server, but data synchronization and long latency will be foreseeable issues due to the possible significant physical distance and transmission delays between BSs and the central server. Therefore, the distributed framework for the communication system still needs further discussion (***Challenge 2***).

Another practical issue mentioned in ***Challenge 3*** is that a pair of trained DNN models cannot handle the dynamic circumstances in wireless networks. Observing (6.60), (6.63), and (6.64), it can be confirmed that the dimensions of the state and action spaces (the scale of the input and the output layers of the DNN model) are jointly determined by multiple parameters, such as the number of antennas, STAR elements, and active users served by the STARS. Although the numbers of antennas and STARS elements are generally constants, the number of connected users is likely to be a time-varying variable. Therefore, when the number of active users in the cell changes, the old policy may no longer be applicable due to changes in the state and action space. In this case, as mentioned in ***Challenge 3***, it needs additional time and exploration to train a pair of new models. During the training process, since naive models are employed, mismatched estimations are likely to be produced, resulting in degraded beamforming and a consequential decline in the SINR. Therefore, in order to provide a comprehensive distributed solution for the BSs and to resolve ***Challenge 2*** and ***Challenge 3***, the AFFL framework is proposed.

1. *Generalized Framework*: The generalized framework of the AFFL is shown in Fig. 6.13. BSs are allocated into different federations according to the number of active users associated with the BSs, which guarantees that the model $\omega_{b,t}^{\kappa}$ of the agents in federation κ has the same specifications and they can share the global model $\omega_{G,t}^{\kappa}$, where κ represents the federation. Since the BSs are likely to have a diverse number of active users, thus, the agent in each cell needs different scales of input layer, hidden layer, and output layers, and we cannot federate all agents with different scales of DNN models via a single federation. As shown in Fig. 6.13, agents have to form several federations, which are indexed by $1 \ldots f$, and the global model of each federation has to be uploaded and stored at the central server. We apply the online training for each agent, and the agents in the same federation use federated averaging to achieve parameter exchange. Since the BSs are linked by wire and fiber, this model does not incur huge spectrum overhead like applying FL among terminal devices. With the assistance of FL, we

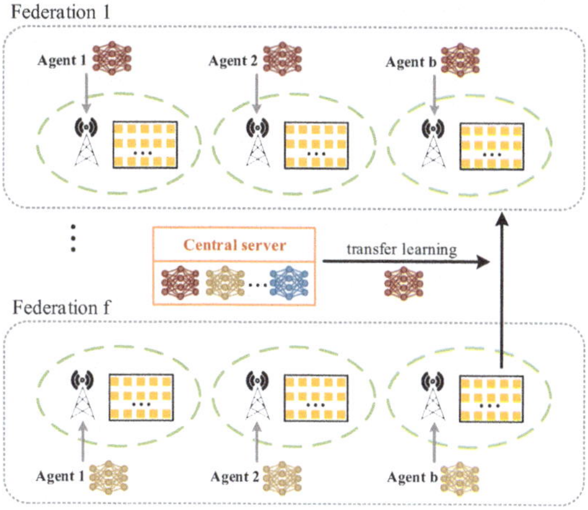

Fig. 6.13 Performances of different algorithms for STARS

provide a solution for ***Challenge 2***, efficiently utilizing the extensive transition samples and hardware resources from different BSs to accelerate training.

To address ***Challenge 3***, we employ the TL in the AFFL framework to enable a join and quit mechanism for federations. In classic FL frameworks, the federation is assumed to be unchanged during the learning process. As mentioned above, agents in the communication system need to change models and be sent to another federation once the number of users in the cell varies. Before the agent changes its federation, the central server has to issue the agent the global model of the target federation as a smart initialization, and then the agents can provide high-quality decisions for users immediately. Since the agent can flow freely among multiple federations, we name the framework AFFL. We will explain the technical details of the federated aggregation and TL, respectively.

2. *Federated Aggregation*: For PPO agents in the same federation, their training process is divided into four stages loops, which are local model update, local model upload, global model update, and globe model download:

- Local model update: Each agent updates the actor $\omega^{\kappa}_{a,b}$ and critic $\omega^{\kappa}_{c,b}$ locally according to the principle of PPO2. The models are time-varying during the training, which can also be denoted as $\omega^{\kappa}_{a,b,t}$ and critic $\omega^{\kappa}_{c,b,t}$.
- Local model upload: After an interval of training, each agent has to send its models $\omega^{\kappa}_{a,b,t}$ and critic $\omega^{\kappa}_{c,b,t}$ to the central server. Thus the agent only needs to save its own models, but the central server has access to all models.
- Global model update: After receiving models from all agents that belong to federation κ, the globe models of κ can be generated by averaging the parameters of local models, which can be expressed as

$$\omega^{\kappa}_{a,G,t} = \frac{1}{B^{\kappa}} \sum_{b^{\kappa}=1}^{B^{\kappa}} \omega^{\kappa}_{a,b,t}, \tag{6.75}$$

$$\omega_{c,G,t}^{\kappa} = \frac{1}{B^{\kappa}} \sum_{b^{\kappa}=1}^{B^{\kappa}} \omega_{c,b,t}^{\kappa}, \tag{6.76}$$

where B^{κ} represents the number of BSs in κ.

- Globe model download: Finally, each agent downloads the globe critic model to replace its local critic model

$$\omega_{c,b,t}^{\kappa} = \omega_{c,G,t}^{\kappa}, \forall b. \tag{6.77}$$

Please note that the local actor model cannot be updated by the globe actor model since the CSI and user distribution are not identical in each cell.

This mode enables distributed learning and avoids frequent data transfers compared to the DPPO framework since the frequency of federated aggregation can be adjusted.

3. *Transfer Federation*: When the agent b^{κ} transfers form federation $\kappa = 0$ to $\kappa = k$, where k is any nonzero integer, it no longer participates in federated aggregation of federation 0 in the upcoming time slot. Before the agent b^0 transferred to b^k, the central server needs to transfer models to b to replace the previous local models, which can be expressed as

$$\omega_{a,b,t}^{\kappa} = \omega_{c,G,t}^{k}, \tag{6.78}$$

$$\omega_{c,b,t}^{\kappa} = \omega_{c,G,t}^{k}. \tag{6.79}$$

Combining all the above steps, the pseudocode of the AFFL-PPO algorithm for federation κ is given in Algorithm 7. Please note that the given pseudocode is for one federation for a clear presentation, and in practice multiple federations are working in parallel.

Remark 6.5 Thanks to the global model provided by TL, on the one hand, the new joined agent owns a trained model, on the other hand, it does not need to have a large exploration rate as a fresh agent, and thus it can keep a small action variance as a well-trained agent. Thus, there is only a negligible communication performance degradation for the agent who transferred federation.

6.3.3 Numerical Results

In the following, we will present the simulation results to reveal the performance of the proposed STARS-aided NOMA network and the ML algorithm. Meanwhile, by analyzing the numerical results, we verify the provided solution and insights for ***Challenge 1–3***. We will first discuss the performance of the aforementioned two STARS models and the partition approaches, and then we verify the performance

Algorithm 7 AFFL-PPO algorithm for the sum rate optimization

for each federation κ **do**
  Initialization: Determine the DNN specifications based on the number of active users and STARS elements.
  Initialize the actor network $\omega^{\kappa}_{a,b,0}$, critic network $\omega^{\kappa}_{c,b,0}$, and global model $\omega^{\kappa}_{a,G,0}$, $\omega^{\kappa}_{c,G,0}$.
  for each BS b^{κ} **do**
    for each iteration **do**
      if globe update = Ture **then**
        Update globe model $\omega^{\kappa}_{a,G,t}$, $\omega^{\kappa}_{c,G,t}$ and local critic models $\omega^{\kappa}_{c,b,t}$ according to (6.75) (6.76) (6.77).
      end if
      if new agent b_0 joining in = True **then**
        Initialize models $\omega^{\kappa}_{a,b_0,t}$ and $\omega^{\kappa}_{c,b_0,t}$ according to (6.78) (6.79).
      end if
      for each time step t **do**
        Observe $\boldsymbol{s}_t$ and choose $\boldsymbol{a}_t$ according to the policy $\pi_{\omega_{\text{old}}}(\boldsymbol{a}_t|\boldsymbol{s}_t)$.
        Execute action $\boldsymbol{a}_t$, observe r_t and $\boldsymbol{s}_{t+1}$.
        Record transition $(\boldsymbol{s}_t, \boldsymbol{a}, r_t, \boldsymbol{s}_{t+1})$.
        Calculate advantage estimates according to (6.67).
      end for
      for each training epoch **do**
        Random sample and calculate loss according to (6.73).
        Calculate the policy gradient $\nabla_{\omega} L_t(\omega)$ and train actor/critic networks according to (6.74).
      end for
    end for
  end for
end for

of the proposed AFFL model by comparing the performance of multiagent AFFL training and single-cell training.

6.3.3.1 Simulation Setup

In the simulation, the STARS located 4km away from the BS,[4] where the users associated with the STARS are randomly distributed around the STARS, and the users are assumed to have random moving during each time step. There are 30 time steps in the transmission duration, and we assume that the CSI varies per each time step. It is worth noting that the special relationship between the BS and the RIS is the same in different cellulars, but the realization of the user distribution, user movements, as well as channel fading blocks are independent in each cell. For the DRL agent, both the actor and the critic network have four-layer structure, where the

[4] A larger BS to STARS distance is adopted as users at the edge of the cell eager to signal enhancements, and users with closer distances can obtain a satisfactory data rate with a direct link.

Table 6.2 Default parameters

Parameter	Description	Value
M	The number of tiles	6
N	Total element number	60
f_c	Central carrier frequency	5 GHz
W	Bandwidth	1 MHz
K	Users per cell	4
P_{max}	Maximum power per antenna	29 dBm
K_{AR}, K_{RU}	Rician factors	3 dB
σ	Noise power density	−95.2 dBm/MHz
α	Learning rate	3×10^{-4}
γ	Discount factor	0.99
e	Batch size	80 samples
N/A	Optimizer	Adam

"tanh" function is employed for the hidden layer. The scale of the DNN is adjusted according to the environment, and the presented simulation results are all given by the agent with empirically optimal parameters. A decaying action noise is invoked to balance the exploration and the exploitation, which starts at 0.6 at the beginning of the training, linearly decays, and finally stays at 0.05. The rest of the default parameters are listed in Table 6.2.

6.3.3.2 STARS Partition

Figure 6.14 demonstrates a pair of examples of the partitioned STARS in the same fading block, where Fig. 6.14a is an example of the ES protocol, and Fig. 6.14b is an example of the MS protocol. Each square represents a STARS element, and there are 60 elements on the STARS in the example. The elements with the same number and color label are allocated into the same tile. It is worth noting that, in the simulation of the MS protocol, tiles 1–3 work in the R mode, while tiles 4–6 are employed to serve T users. It can be observed that the partition given by the agent has significant differences for ES and MS models under the same fading block. The partition given for the MS-based STARS can be explained by the classical greedy algorithm and the NOMA gain theory. However, the partition for the ES-based STARS produces two large tiles, tile 3 and tile 6. The reason leading to this phenomenon may be the QoS constraint. Since the ES protocol cannot adjust the phase arbitrarily, it is harder to meet the QoS requirements with small-scale tiles, such as tiles 1 and 5 in Fig. 6.14b. Hence, the reason for having two larger tiles in Fig. 6.14a may not be that the two major tiles are needed, but to avoid the appearance of very small tiles. Therefore, although the interpretability of ML methods is relatively deficient, the partition results given by the DRL agent are interpretable with the help of theory in communication.

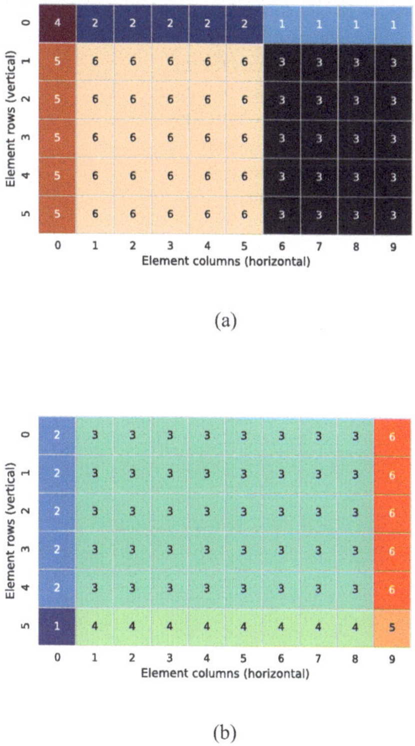

Fig. 6.14 Example of the STARS partition. (**a**) Example partitioning for the ES STARS. (**b**) Example partitioning for the MS STARS

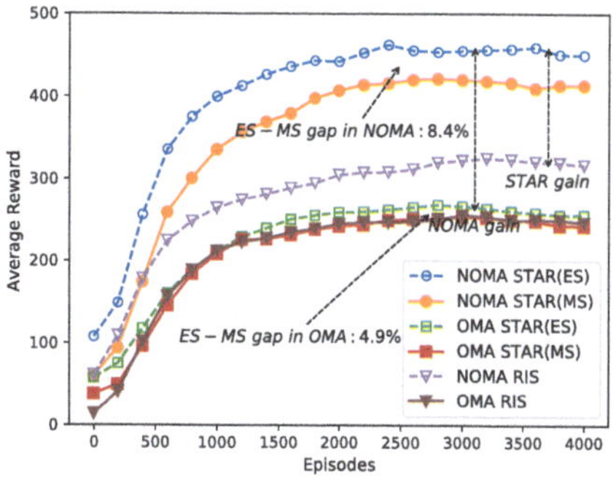

Fig. 6.15 Performances of different types for STARS in OMA/NOMA networks

Figure 6.15 plots the averaged reward[5] obtained during the simulated duration of 3 STARS-assisted NOMA cells to reveal the performance of the STARS. Significant NOMA gains and STARS gains can be observed Fig. 6.15 aside from which type of STARS is employed. In the NOMA networks, the STARSs obtain

[5] The average reward and average data rate refer to the average value of the reward and data rate obtained by the agent of multiple BSs and episodes, which is invoked to reduce the effect caused by the random action explorations and channel realizations.

Fig. 6.16 Performance comparison of different partitioning schemes for the STARS. (**a**) Reward obtained by different TRC controlling schemes. (**b**) Throughput obtained by different TRC controlling schemes

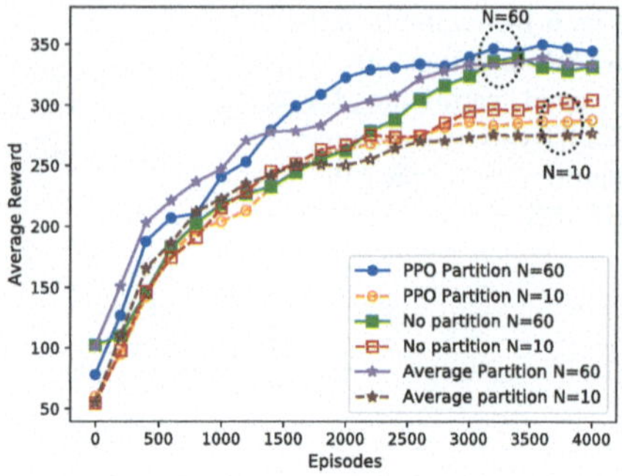

(a)

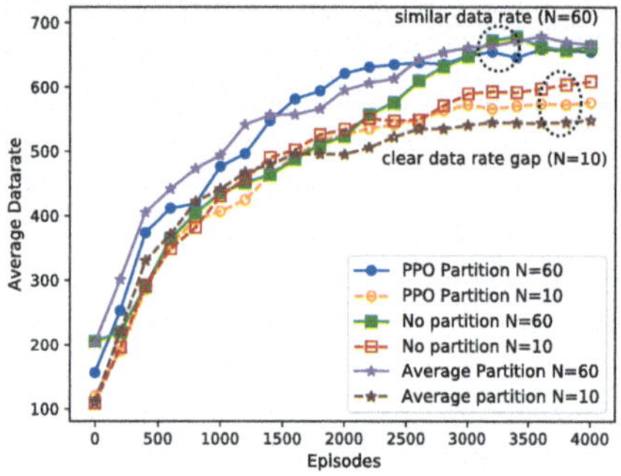

(b)

a reward of around 400 and outperform the conventional RIS investigated in Liu et al. (2022), since the reflecting-only are not capable of severing all users. Furthermore, in both OMA and NOMA cases, the ES protocol achieved superior reward compared to the MS protocol. This result indicates that though ES elements cannot independently configure the TRC, a larger equivalent element number still dominates the performance. It is worth emphasizing that the ES protocol has a stronger affinity to NOMA networks. In the NOMA network, the ES protocol achieved a gain of about 8.4% compared to the MS protocol, while in contrast to the OMA network, only a gain of 4.9% was achieved.

To explain and verify the necessity and advantage of the STARS partition, we compared the system performance of three TRC controlling schemes in Fig. 6.16, namely intelligent partitioning, average partitioning, and element-based intelligent control. Figure 6.16b plots the obtained reward of the three approaches in cases of $N = 10$ and $N = 60$. While only $10(2 \times 5, M = 2)$ elements of STARS are employed, the element-based TRC control scheme achieves the highest reward, as the independent control can provide precise TRCs for each element. However, once a STARS having a large number of elements is invoked, the rank of the reward is changed, where the independent control scheme achieved an inferior reward to the partition-based control, which is caused by the increased element number and

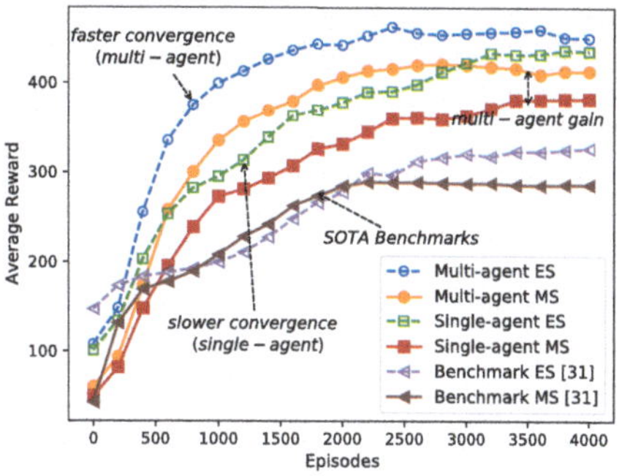

Fig. 6.17 Performances of different learning frameworks for STARS

complexity that degrades the performance of the trained agent as we analyzed in Remark 6.4. This trend is evidence to prove that when a larger scale of STARS is employed, unless accepting an extremely large training cost, the partition-based TRC control is likely to outperform the element-based control scheme, since the STARS partition can significantly simplify the problem to improve the performance of DRL agents. Meanwhile, the PPO-based partition outperformed the average partition in both realizations, which indicates the advantage of the intelligent partition. If we shift the focus to Fig. 6.16b, it can be observed that in the case of $N = 10$, the throughput has exactly the same ranking and trend as the reward, but at $N = 60$ the throughput of the three schemes became nearly coinciding. The results indicate that the intelligent partitioning scheme obtains additional rewards due to meeting the users' QoS requirements, suggesting that the DRL partitioning does not unilaterally pursue data rate maximization like the water-filling algorithm, and coincides with the partitioning example shown in Fig. 6.14a.

6.3.3.3 AFFL Distributed Model

Now we refocus on the validation of the proposed AFFL framework. Since we aim to evaluate the training process and performance of the agents, the performance indicators of the communication system are no longer in our account, and the reward is considered as the sole performance indicator to evaluate the system performance.

Figure 6.17 demonstrates the training process of a 3-cell federation started with random initialized models. In order to validate the performance of the DRL-based beamforming and the AFFL framework, the algorithm in Guo et al. (2021) is employed as a state-of-the-art (SOTA) benchmark. It can be observed that the proposed solutions outperform the benchmark algorithm in terms of convergence and optimality. Comparing two pairs of curves for the ES protocol and the MS protocol, we can firstly observe a significant reward gain in the multiagent case where AFFL is applied. Another significant advantage of AFFL is convergence. Although the large action noise at the early training stage affects the observation of convergence to some extent, taking the ES case as an example, the convergence

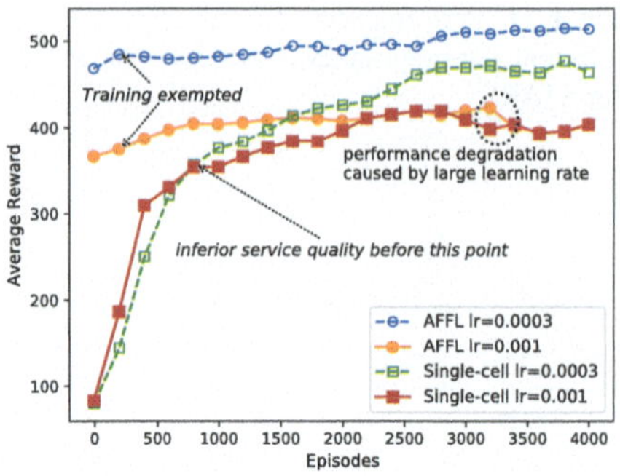

Fig. 6.18 Performances of different learning frameworks for STARS

shown by the blue dashed curve is better than the green one. Therefore, in a limited training period, the AFFL framework achieves better optimality and convergence by aggregating the critic model from multiple agents.

Figure 6.18 plots the training process against different learning rates of a cell that switches federation needs to experience. In this scenario, we assume that a cell becomes a new member of the federation mentioned in Fig. 6.17 due to the change in the number of users, and we plot the reward of the new member agent. In case of the AFFL model is employed, since a fully trained model is distributed to the agent, it can obtain eligible rewards from episode 1, which indicates that users do not have to wait for the training of the agent and can be served immediately. Although the agent still obtains a small reward improvement with training, the initial service quality is still guaranteed as we claimed in Remark 6.5. Conversely, users who are not under the AFFL model are not so lucky, during the first 1000 episodes of the training time, these users are likely to experience poor service due to the immature model. Even if we sacrifice optimality a bit and set a larger learning rate 0.001 to shorten the training process, a wait of about 500 epochs is still necessary. Therefore, this result illustrates that with the aid of TL, the AFFL model is capable of solving ***Challenge 3***.

Table 6.3 presents the training time consumption per agent of different partitioning approaches and training frameworks. It can be observed that the training time consumed by the tile-based beamforming scheme is significantly diminished than the element-based beamforming, and the DRL-based beamforming does not produce discernible complexity compared to the averaged partitioning. In terms of the training framework, by adopting the model aggregation of multiple agents, the AFFL framework greatly reduces the training time of the model. Specifically, for the problem of dynamic network requirements mentioned in ***Challenge 3***, the AFFL model does not require any additional retraining but invokes model transfers to solve the problem.

Table 6.3 Training time consumption of different partitioning approaches and training frameworks

Element number	No partition	Average partition	Individual PPO	AFFL-PPO (initialization)	AFFL-PPO (transfer)
10	41 m 55 s	25 m 16 s	27 m 02 s	17 m 12 s	0 s
60	321 m 08 s	34 m 26 s	34 m 30 s	19 m 39 s	0 s

6.4 Summary

This chapter introduces the employment of ML tools for STARS-aided wireless communications, where hybrid RL algorithms are developed for the STARS beamforming design with the coupled phase-shift model and a distributed DRL algorithm is conceived for STARS-aided downlink multicell NOMA networks.

References

3rd Generation Partnership Project. (2018). Study on 3D channel model for LTE (3GPP TR 36.873 release 12).

Abels, A., Roijers, D., Lenaerts, T., Nowé, A., & Steckelmacher, D. (2019). Dynamic weights in multi-objective deep reinforcement learning. In: *International Conference on Machine Learning* (pp. 11–20). PMLR.

Abraham, I., Prabhakar, A., & Murphey, T. D. (2021). An ergodic measure for active learning from equilibrium. *IEEE Transactions on Automation Science and Engineering, 18*, 917–931.

Arulkumaran, K., Deisenroth, M. P., Brundage, M., & Bharath, A. A. (2017). Deep reinforcement learning: A brief survey. *IEEE Signal Processing Magazine, 34*, 26–38.

Chafaa, I., Negrel, R., Veronica Belmega, E., & Debbah, M. (2022). Self-supervised deep learning for mmwave beam steering exploiting sub-6 GHz channels. *IEEE Transactions on Wireless Communications, 21*(10), 8803–8816.

Delalleau, O., Peter, M., Alonso, E., & Logut, A. (2019). Discrete and continuous action representation for practical RL in video games. arXiv preprint arXiv:1912.11077.

Feriani, A., & Hossain, E. (2021). Single and multi-agent deep reinforcement learning for AI-enabled wireless networks: A tutorial. *IEEE Communications Surveys & Tutorials, 23*, 1226–1252.

Guo, Y., Fang, F., Cai, D., & Ding, Z. (2021). Energy-efficient design for a NOMA assisted STAR-RIS network with deep reinforcement learning. arXiv preprint arXiv:2111.15464.

Heess, N., TB, D., Sriram, S., Lemmon, J., Merel, J., Wayne, G., Tassa, Y., Erez, T., Wang, Z., Eslami, S., et al. (2017). Emergence of locomotion behaviours in rich environments. arXiv preprint arXiv:1707.02286.

Li, B., Tang, H., Zheng, Y., Hao, J., Li, P., Wang, Z., Meng, Z., & Wang, L. (2021a). HyAR: Addressing discrete-continuous action reinforcement learning via hybrid action representation. arXiv preprint arXiv:2109.05490.

Li, C., Xia, J., Liu, F., Li, D., Fan, L., Karagiannidis, G. K., & Nallanathan, A. (2021b). Dynamic offloading for multiuser muti-CAP MEC networks: A deep reinforcement learning approach. *IEEE Transactions on Vehicular Technology, 70*, 2922–2927.

Li, Y., Aghvami, A. H., & Deng, Y. (2021c). Joint resource block and beamforming optimization for cellular-connected UAV networks: A hybrid D3QN-DDPG approach. arXiv preprint arXiv:2102.13222.

Lillicrap, T. P., Hunt, J. J., Pritzel, A., Heess, N., Erez, T., Tassa, Y., Silver, D., & Wierstra, D. (2015). Continuous control with deep reinforcement learning. arXiv preprint arXiv:1509.02971.

Liu, Y., Mu, X., Liu, X., Di Renzo, M., Ding, Z., & Schober, R. (2022). Reconfigurable intelligent surface-aided multi-user networks: Interplay between NOMA and RIS. *IEEE Wireless Communications, 29*, 169–176.

Najafi, M., Jamali, V., Schober, R., & Poor, H. V. (2021). Physics-based modeling and scalable optimization of large intelligent reflecting surfaces. *IEEE Transactions on Communications, 69*, 2673–2691.

Naughton, N., Sun, J., Tekinalp, A., Parthasarathy, T., Chowdhary, G., & Gazzola, M. (2021). Elastica: A compliant mechanics environment for soft robotic control. *IEEE Robotics and Automation Letters, 6*, 3389–3396.

Neunert, M., Abdolmaleki, A., Wulfmeier, M., Lampe, T., Springenberg, T., Hafner, R., Romano, F., Buchli, J., Heess, N., & Riedmiller, M. (2020). Continuous-discrete reinforcement learning for hybrid control in robotics. In: *Proceedings of the CoRL* (Vol. 100, pp. 735–751).

Qiu, C., Hu, Y., Chen, Y., & Zeng, B. (2019). Deep deterministic policy gradient (DDPG)-based energy harvesting wireless communications. *IEEE Internet of Things Journal, 6*, 8577–8588.

Schaul, T., Horgan, D., Gregor, K., & Silver, D. (2015). Universal value function approximators. In: *International Conference on Machine Learning* (pp. 1312–1320). PMLR.

Schulman, J., Wolski, F., Dhariwal, P., Radford, A., & Klimov, O. (2017). Proximal policy optimization algorithms. arXiv preprint arXiv:1707.06347.

Yang, Z., Xu, W., Pan, C., Pan, Y., & Chen, M. (2017). On the optimality of power allocation for NOMA downlinks with individual QoS constraints. *IEEE Communications Letters, 21*, 1649–1652.

Yu, X., Jamali, V., Xu, D., Ng, D. W. K., & Schober, R. (2021). Smart and reconfigurable wireless communications: From IRS modeling to algorithm design. *IEEE Wireless Communications, 28*, 118–125.

Zhang, S., & Zhang, R. (2020). Capacity characterization for intelligent reflecting surface aided MIMO communication. *IEEE Journal on Selected Areas in Communications, 38*, 1823–1838.

Zuo, J., Liu, Y., Ding, Z., Song, L., & Vincent Poor, H. (2022). Joint design for simultaneously transmitting and reflecting (STAR) RIS assisted NOMA systems. *IEEE Transactions on Wireless Communications, 22*(1), 611–626

Chapter 7
STARS-Aided Integrated Sensing and Communications

Abstract In this chapter, we will introduce the employment of STARS for supporting integrated sensing and communications (ISACs) in future wireless networks. In particular, a sensing-at-STARS structure is proposed with numerical case studies.

7.1 Sensing-at-STARS

Next-generation wireless networks are envisioned to support not only communication but also sensing functions, which has made the concept of ISAC a hot research topic recently (Liu et al., 2022). The key objective of ISAC is to carry out communication and sensing (C&S) functions using a shared hardware platform and signals. However, these two functions have different requirements for system design. For example, the communication function can effectively exploit the non-LoS path for data delivery, while the sensing function generally relies on the LoS path for target detection and parameter estimation. In practice, the LoS path does not always exist due to the blockage, which makes target sensing quite challenging. RISs are a promising solution to address this issue by establishing a virtual LoS path from the BS to the target (Buzzi et al., 2022). However, employing conventional reflecting/transmitting-only RISs has the following two disadvantages. On the one hand, conventional RISs can only achieve half-space coverage for both C&S. On the other hand, in RIS-ISAC systems, the communication users and sensing targets have to be located on the same side of the RIS and share the same passive beamforming matrix, which restricts design flexibility. Compared to conventional RISs, STARSs provide a paradigm shift for ISAC design. In the following, we first introduce some basic system configurations of STARS-ISAC, which is followed by an introduction of a promising *sensing-at-STARS* architecture.

As shown in Fig. 7.1, based on the full-space coverage and signal splitting nature, there are two possible configurations of STARS-ISAC, namely integrated full-space (IFS) configuration (Zhang et al., 5795; Wang et al., 2023a) and separated half-space (SHS) configuration (Liu et al., 2023; Wang et al., 2023,b; Sun et al., 2023; Meng et al., 2024). In the following, we elaborate on these configurations:

Y. Liu et al., *Simultaneously Transmitting and Reflecting Surfaces for Wireless Communications*, Wireless Networks, https://doi.org/10.1007/978-3-031-76848-4_7

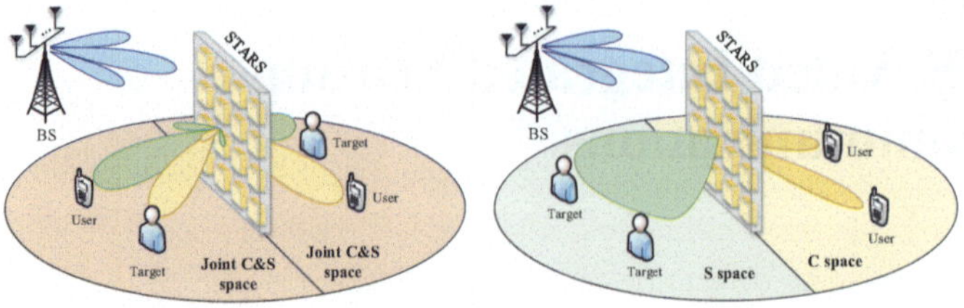

(a) (b)

Fig. 7.1 Illustration of two systems' configurations for STARS-ISAC systems. (**a**) Integrated full-space configuration. (**b**) Separated half-space configuration

- **Integrated Full-Space Configuration:** As illustrated in Fig. 7.1a, in the IFS configuration, communication users and sensing targets exist on both sides of STARS. This configuration enables 360° coverage and full DoFs for both C&S functions. Drawing on this configuration, Zhang et al. (5795) crafted a STARS beamforming design based on the TS protocol, which minimized the Cramér-Rao bound (CRB) for target sensing while ensuring the QoS requirements of uplink communications. In another study, Wang et al. (2023a) proposed an ES-based STARS beamforming design, aiming to maximize the signal-to-clutter-plus-noise ratio (SCNR) for sensing under the constraint of minimum QoS requirements for downlink communications. While the IFS configuration can achieve full DoFs for ISAC, it is important to note that the joint C&S beamforming design for STARS can lead to substantial computational complexity that may hinder practical implementation, especially considering the inherently passive nature of STARS.
- **Separated Half-Space Configuration:** In contrast to the IFS configuration, the SHS configuration separates the entire space into two separated spaces (termed as S and C spaces) for C&S, respectively, as depicted in Fig. 7.1b. The SHS configuration is a new possibility introduced by STARS compared to conventional RISs, where two generally independent passive STARS beamforming matrices can be tailored for C&S, respectively. This is important for a flexible and low-complexity ISAC design since the C&S functions generally have distinct beamforming requirements, e.g., wide beams for sensing and narrow beams for communication in Fig. 7.1b. Building on this model, Liu et al. (2023) further studied SNR maximization for target sensing. As a further advance, NOMA was exploited in Wang et al. (2023) to mitigate both sensing-to-communication interference and inter-user interference in SHS-based STARS-ISAC systems. The physical layer security (PLS) issue in SHS-based STARS-ISAC systems was explored in Wang et al. (2023b) and Sun et al. (2023), with a focus on independent and coupled phase-shift models, respectively. In a separate study, Meng et al. (2024) conceived a novel SHS-based sensing-assisted communication framework for vehicular networks. In this framework, STARS are mounted on vehicles to enhance the sensing echoes to the BS through the reflection

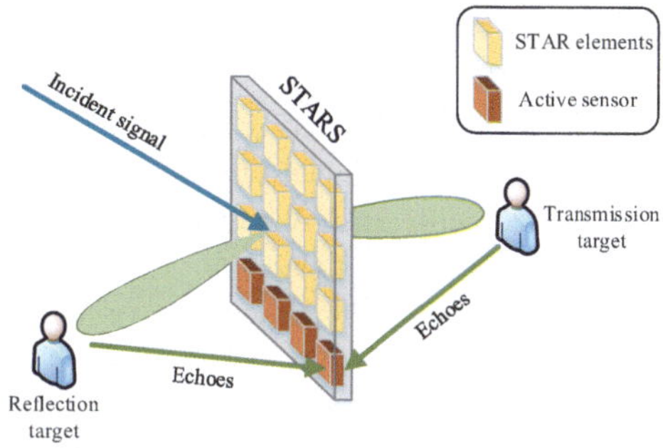

Fig. 7.2 Illustration of the sensing-at-STARS architecture

beamforming and improve the communication performance inside the vehicle through the transmission beamforming.

While STARSs provide notable advantages for ISAC designs, it also introduces specific challenges, particularly for sensing. Firstly, akin to conventional RIS-ISAC (Shao et al., 2022), STARS-ISAC exhibits the issue of multi-hop path loss. When carrying out the sensing function at the BS, the sensing echoes experience substantial path loss through a cascaded BS→STARS→target→STARS→BS path. Secondly, due to the inherent energy-splitting effect, there is a power leakage issue at STARSs that further diminishes the echo power. Lastly, the mixing of sensing echoes from both transmission and reflection sides in the STARS→BS link complicates the accurate estimation of target locations.

To overcome these challenges, we proposed a *sensing-at-STARS* architecture, as shown in Fig. 7.2. This architecture, where dedicated active sensors are installed on STARS, presents the following benefits:

- **Enhanced Echo Power:** By processing echoes directly at STARS rather than at BSs, this architecture significantly reduces multi-hop path loss and mitigates power leakage issues at STARS, thereby substantially boosting echo power and improving sensing performance.
- **Reduced Echo Aliasing:** The mix of echoes from transmission and reflection sides in the STARS→BS link is eliminated. Instead, active sensors on STARS directly receive these echoes, thus facilitating easier and more accurate target location estimations.
- **Low-Complexity Beamforming:** The sensing-at-BS architecture requires consideration of both STARS→target and STARS→BS links for beamforming design, leading to high complexity. By contrast, the sensing-at-STARS architecture allows beamforming to be focused solely on enhancing the STARS→target link, thereby simplifying the beamforming design.

In the following, we will provide a case study for characterizing the performance of the proposed sensing-at-STARS architecture.

7.2 System Model

As shown in Fig. 7.3, we consider a narrowband STARS-enabled ISAC system, where the BS is equipped with a ULA consisting of M antennas and the STARS is equipped with a UPA consisting of N passive transmission-reflection (T-R) elements denoted by the set $\mathcal{N}$. The whole space is divided into two half-spaces by STARS, namely the *sensing space* and the *communication space*. Without loss of generality, we assume that the sensing space is on the reflection side and the communication space is on the transmission side. There is a single sensing target of interest in the sensing space and K single-antenna communication users, whose indices are collected in $\mathcal{K}$, in the communication space. The direct links between BS and target/users are assumed to be blocked. To tackle the severe path loss, we propose a sensing-at-STARS structure, where a dedicated low-cost sensor with a ULA consisting of N_s elements is mounted on STARS. The side of the sensor facing the communication space is physically blocked to avoid the clutter signal from the communication space. Furthermore, we consider a coherent time block of length L, during which the communication channels and sensing target parameters remain approximately constant.

7.2.1 Signal Model

The ES model is exploited to support simultaneous transmission and reflection of the STARS. In particular, the incident signal at the STARS from the BS is split into the sensing signal in the sensing space and the communication signal in the communication space. Denote $\mathbf{\Theta}_t \in \mathbb{C}^{N \times N}$ and $\mathbf{\Theta}_r \in \mathbb{C}^{N \times N}$ as the matrices of the TCs and RCs, respectively, which can be modeled as

$$\mathbf{\Theta}_i = \mathrm{diag}\left(\beta_{i,1} e^{j\varphi_{i,1}}, \ldots, \beta_{i,N} e^{j\varphi_{i,N}}\right), \forall i \in \{t, r\}. \tag{7.1}$$

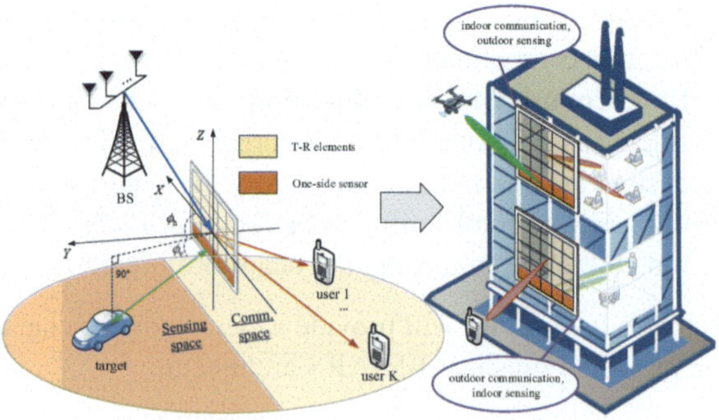

Fig. 7.3 Illustration of the STARS-enabled ISAC system

In the above expression, $\beta_{i,n} \in [0, 1]$ and $\varphi_{i,n} \in [0, 2\pi]$ denote the amplitude and phase-shift response of the n-th element. To fully investigate the role of STARS in the proposed system, both independent and coupled phase-shift models are considered in this chapter.

To carry out sensing and communication simultaneously, the BS transmits the following joint signal at the time index t:

$$\mathbf{x}(t) = \mathbf{P}\mathbf{c}(t) + \mathbf{s}(t) = \sum_{k \in \mathcal{K}} \mathbf{p}_k c_k(t) + \mathbf{s}(t), \tag{7.2}$$

where $\mathbf{P} = [\mathbf{p}_1, \ldots, \mathbf{p}_K] \in \mathbb{C}^{M \times K}$ denotes the transmit beamforming matrix for delivering the information streams $\mathbf{c}(t) = [c_1(t), \ldots, c_K(t)]^T \in \mathbb{C}^{K \times 1}$ to the K communication users. The signal, $\mathbf{s}(t) \in \mathbb{C}^{M \times 1}$, is the dedicated sensing signal to achieve the full DoFs for target sensing (Liu et al., 2020). The multiple beam transmission is exploited by the dedicated sensing signal. Thus, the covariance matrix of it, i.e., $\mathbf{R}_s = \mathbb{E}[\mathbf{s}(t)\mathbf{s}^H(t)]$, has a general rank. Furthermore, the communication signals are modeled as independent Gaussian random signals with zero mean and unit power, and the dedicated sensing signal is generated by pseudorandom coding, so that $\mathbb{E}[\mathbf{c}(t)\mathbf{c}(t)^H] = \mathbf{I}_K$ and $\mathbb{E}[\mathbf{c}(t)\mathbf{s}(t)^H] = \mathbf{0}_{K \times M}$. Thus, the covariance matrix of the transmit signal $\mathbf{x}(t)$ is given by

$$\mathbf{R}_x = \mathbb{E}[\mathbf{x}(t)\mathbf{x}(t)^H] = \mathbf{P}\mathbf{P}^H + \mathbf{R}_s. \tag{7.3}$$

In practice, $\mathbf{R}_x$ can be calculated by averaging over L time indexes as follows:

$$\mathbf{R}_x \approx \frac{1}{L}\mathbf{X}\mathbf{X}^H, \tag{7.4}$$

where $\mathbf{X} = [\mathbf{x}(1), \ldots, \mathbf{x}(L)]$. The above approximation is accurate when L is large. Thus, we assume the accurate equality of (7.4) is achieved throughout this chapter.

Given the transmit signal at the BS, the received signal at the k-th communication user in the communication space can be modeled as[1]

$$y_{c,k}(t) = \underbrace{\mathbf{h}_k^H \mathbf{\Theta}_t \mathbf{G}\mathbf{p}_k c_k(t)}_{\text{desired signal}} + \underbrace{\sum_{i \in \mathcal{K} \setminus k} \mathbf{h}_k^H \mathbf{\Theta}_t \mathbf{G}\mathbf{p}_i c_i(t)}_{\text{inter-user interference}} + \underbrace{\mathbf{h}_k^H \mathbf{\Theta}_t \mathbf{G}\mathbf{s}(t)}_{\text{sensing interference}} + n_k(t), \tag{7.5}$$

where $\mathbf{G} \in \mathbb{C}^{N \times M}$ denotes the baseband channel matrix between the BS and the STARS, $\mathbf{h}_k \in \mathbb{C}^{N \times 1}$ denotes the baseband channel vector between the STARS and user k, and $n_k \sim \mathcal{CN}(0, \sigma_k^2)$ denotes the AWGN at user k with the variance σ_k^2.

[1] Here, we omit the interference caused by the echo signal from the sensing space because its power at the communication user is negligible.

Then, in the sensing space, the received echo signal from the target over a coherent time block of length L at the sensors can be modeled as

$$\mathbf{Y}_s = \alpha \mathbf{b}(\phi_h, \phi_v)\mathbf{a}^T(\phi_h, \phi_v)\mathbf{\Theta}_r \mathbf{G}\mathbf{X} + \mathbf{N}_s, \tag{7.6}$$

where $\alpha \in \mathbb{C}$ is the complex amplitude determined by the round-trip path loss and the complex reflection factor of the target, ϕ_h and ϕ_v denote azimuth and elevation direction of arrivals (DOAs) of the target with respect to the STARS, respectively, $\mathbf{a}(\phi_h, \phi_v) \in \mathbb{C}^{N\times 1}$ denotes the steering vector of the STARS, $\mathbf{b}(\phi_h, \phi_v) \in \mathbb{C}^{N_s\times 1}$ denotes the steering vector of the sensors, and $\mathbf{N}_s$ denotes the AWGN noise with each entry obeying $\mathcal{CN}(0, \sigma_s^2)$. According to Manikas (2004), the steering vectors under the planar wave assumption can be modeled as

$$\mathbf{a}(\phi_h, \phi_v) = \exp(-j[\mathbf{r}_X, \mathbf{r}_Y, \mathbf{r}_Z]\mathbf{k}(\phi_h, \phi_v)), \tag{7.7}$$

$$\mathbf{b}(\phi_h, \phi_v) = \exp(-j[\bar{\mathbf{r}}_X, \bar{\mathbf{r}}_Y, \bar{\mathbf{r}}_Z]\mathbf{k}(\phi_h, \phi_v)), \tag{7.8}$$

where $[\mathbf{r}_X, \mathbf{r}_Y, \mathbf{r}_Z] \in \mathbb{R}^{N\times 3}$ and $[\bar{\mathbf{r}}_X, \bar{\mathbf{r}}_Y, \bar{\mathbf{r}}_Z] \in \mathbb{R}^{N_s\times 3}$ have rows representing the Cartesian coordinates of the STARS elements and the sensors. The vector $\mathbf{k}(\phi_h, \phi_v) \in \mathbb{R}^{3\times 1}$ is the wavenumber vector defined as follows:

$$\mathbf{k}(\phi_h, \phi_v) = \frac{2\pi}{\lambda_c}[\cos\phi_h \cos\phi_v, \sin\phi_h \cos\phi_v, \sin\phi_v]^T, \tag{7.9}$$

where λ_c denotes the wavelength of the carrier signal. Without loss of generality, we assume that the ULA array of the sensors is deployed along the X-axis and the UPA array of the STARS is deployed within the (X, Z) plane, i.e., $\bar{\mathbf{r}}_Y = \bar{\mathbf{r}}_Z = \mathbf{0}$ and $\mathbf{r}_Y = \mathbf{0}$. As a consequence, the steering vectors can be simplified as follows:

$$\mathbf{a}(\phi_h, \phi_v) = \exp\Big(-j\frac{2\pi}{\lambda_c}(\mathbf{r}_X \cos\phi_h \cos\phi_v + \mathbf{r}_Z \sin\phi_v)\Big), \tag{7.10}$$

$$\mathbf{b}(\phi_h, \phi_v) = \exp\Big(-j\frac{2\pi}{\lambda_c}\bar{\mathbf{r}}_X \cos\phi_h \cos\phi_v\Big). \tag{7.11}$$

The echo signal model in (7.6) is assumed to be acquired in a specific range-Doppler bin. Thus, the range and Doppler parameters are omitted. Furthermore, in addition to the echo signal from the target, the sensors also receive the interference signal through the direct BS-sensor link. However, in practice, such interference can be effectively eliminated by the offline training (Shao et al., 2022) and thereby is also omitted.

7.2.2 Performance Metrics for Communication and 2D Sensing

A common measure of multiuser communication is the SINR, which determines the achievable rate. Therefore, we focus on optimizing the SINR for each communication user. According to (7.5), the SINR for decoding the desired signal at user k is given by

$$\gamma_k = \frac{|\mathbf{h}_k^H \boldsymbol{\Theta}_t \mathbf{G}\mathbf{p}_k|^2}{\sum_{i \in \mathcal{K}\setminus k} |\mathbf{h}_k^H \boldsymbol{\Theta}_t \mathbf{G}\mathbf{p}_i|^2 + \mathbf{h}_k^H \boldsymbol{\Theta}_t \mathbf{G}\mathbf{R}_s \mathbf{G}^H \boldsymbol{\Theta}_t^H \mathbf{h}_k + \sigma_k^2}. \tag{7.12}$$

For target sensing, we are interested in estimating the 2D DOAs ϕ_h and ϕ_v based on the observations $\mathbf{Y}_s$ over a coherent time block, which can be achieved by the classic maximum likelihood estimation (MLE) presented in Appendix G. After obtaining the estimated $\hat{\phi}_h$ and $\hat{\phi}_v$, the MSE, i.e., $\epsilon_h^2 = \mathbb{E}[|\phi_h - \hat{\phi}_h|^2]$ and $\epsilon_v^2 = \mathbb{E}[|\phi_v - \hat{\phi}_v|^2]$, is commonly used to evaluate the estimation performance. However, the optimization of MSE is intractable since it is difficult to obtain closed-form expressions. Therefore, we are interested in optimizing the CRB for estimating the 2D DOAs ϕ_h and ϕ_v, which provides a lower bound for the MSE and has the closed-form expression. To derive the CRB, we first vectorize the matrix $\mathbf{Y}_s$ as follows:

$$\mathbf{y}_s = \mathrm{vec}(\mathbf{Y}_s) = \mathbf{u} + \mathbf{n}_s, \tag{7.13}$$

where $\mathbf{u} = \mathrm{vec}(\alpha \mathbf{b}(\phi_h, \phi_v)\mathbf{a}^T(\phi_h, \phi_v)\boldsymbol{\Theta}_r \mathbf{G}\mathbf{X})$ and $\mathbf{n}_s = \mathrm{vec}(\mathbf{N}_s)$. Denote $\boldsymbol{\xi} = [\boldsymbol{\phi}^T, \tilde{\boldsymbol{\alpha}}^T]^T$, where $\boldsymbol{\phi} = [\phi_h, \phi_v]^T$ and $\tilde{\boldsymbol{\alpha}} = [\mathrm{Re}(\alpha), \mathrm{Im}(\alpha)]^T$, as the unknown parameters to estimate. Then, the Fisher information matrix (FIM) for estimating the vector $\boldsymbol{\xi}$ can be partitioned as

$$\mathbf{J}_{\xi} = \begin{bmatrix} \mathbf{J}_{\phi\phi} & \mathbf{J}_{\phi\tilde{\alpha}} \\ \mathbf{J}_{\phi\tilde{\alpha}}^T & \mathbf{J}_{\tilde{\alpha}\tilde{\alpha}} \end{bmatrix}. \tag{7.14}$$

Then, the CRB matrix for estimating $\boldsymbol{\phi}$ is given by Bekkerman and Tabrikian (2006)

$$\mathrm{CRB}(\boldsymbol{\phi}) = \left[\mathbf{J}_{\phi\phi} - \mathbf{J}_{\phi\tilde{\alpha}} \mathbf{J}_{\tilde{\alpha}\tilde{\alpha}}^{-1} \mathbf{J}_{\phi\tilde{\alpha}}^T\right]^{-1}, \tag{7.15}$$

where the expressions of the matrices $\mathbf{J}_{\phi\phi}$, $\mathbf{J}_{\phi\tilde{\alpha}}$, and $\mathbf{J}_{\tilde{\alpha}\tilde{\alpha}}$ are derived in Appendix H. It can be observed that $\mathrm{CRB}(\boldsymbol{\phi})$ depends on the target DOAs $\boldsymbol{\phi}$. The knowledge of $\boldsymbol{\phi}$ is thus required to design the joint waveform at the BS and the TCs/RCs at the STARS. Note that, in practice, the target DOAs typically do not change significantly between two adjacent coherent time blocks. Therefore, the estimated or predicted DOAs from estimation results in previous coherent time blocks are sufficient for system design (Liu et al., 2021). Based on the above analysis, we assume that the

DOAs $\boldsymbol{\phi}$ are fixed for the optimization problem. Furthermore, it is assumed that the communication channels of both independent and coupled phase-shift STARS have been obtained by the advanced channel estimation method (Wu et al., 2022).

7.3 STARS-ISAC: CRB Optimization with the Independent Phase-Shift Model

In this section, based on the proposed framework, we focus on the minimization of CRB in the case of STARS with independent phase shifts, which can be simplified to the optimization of a modified FIM. We then propose to solve the resulting optimization problem by invoking the PDD framework.

7.3.1 Problem Formulation

The diagonal elements of the CRB matrix represent the minimum variance of each target parameter estimated by the unbiased estimator (Kay, 1993). Therefore, we aim to minimize the trace of the CRB matrix, while guaranteeing the minimum SINR level at each communication user. To optimize the complicated CRB matrix $\mathrm{CRB}(\boldsymbol{\phi})$, we first transform it into a more tractable form according to the following proposition.

Proposition 7.1 *Minimizing the trace of CRB matrix* $\mathrm{CRB}(\boldsymbol{\phi})$ *is equivalent to solving the following optimization problem:*

$$\min_{\mathbf{U},\boldsymbol{\Omega}} \quad \mathrm{tr}(\mathbf{U}^{-1}) \tag{7.16a}$$

$$\text{s.t.} \quad \begin{bmatrix} \mathbf{J}_{\boldsymbol{\phi}\boldsymbol{\phi}} - \mathbf{U} & \mathbf{J}_{\boldsymbol{\phi}\tilde{\boldsymbol{\alpha}}} \\ \mathbf{J}_{\boldsymbol{\phi}\tilde{\boldsymbol{\alpha}}}^T & \mathbf{J}_{\tilde{\boldsymbol{\alpha}}\tilde{\boldsymbol{\alpha}}} \end{bmatrix} \succeq 0, \tag{7.16b}$$

$$\mathbf{U} \succeq 0, \tag{7.16c}$$

where $\mathbf{U} \in \mathbb{C}^{2\times 2}$ *is an auxiliary matrix and* $\boldsymbol{\Omega}$ *denotes the original optimization variables.*

Proof Firstly, we have that the FIM $\mathbf{J}_{\boldsymbol{\phi}\boldsymbol{\phi}} - \mathbf{J}_{\boldsymbol{\phi}\tilde{\boldsymbol{\alpha}}}\mathbf{J}_{\tilde{\boldsymbol{\alpha}}\tilde{\boldsymbol{\alpha}}}^{-1}\mathbf{J}_{\boldsymbol{\phi}\tilde{\boldsymbol{\alpha}}}^T$ is a positive semidefinite matrix. According to Boyd et al. (2004, Example 3.46), the function $\mathrm{tr}(\mathbf{A}^{-1})$ is matrix decreasing on the positive semidefinite matrix space. Therefore, minimizing $\mathrm{tr}\left([\mathbf{J}_{\boldsymbol{\phi}\boldsymbol{\phi}} - \mathbf{J}_{\boldsymbol{\phi}\tilde{\boldsymbol{\alpha}}}\mathbf{J}_{\tilde{\boldsymbol{\alpha}}\tilde{\boldsymbol{\alpha}}}^{-1}\mathbf{J}_{\boldsymbol{\phi}\tilde{\boldsymbol{\alpha}}}^T]^{-1}\right)$ is equivalent to minimizing $\mathrm{tr}(\mathbf{U}^{-1})$, where $\mathbf{U} \succeq 0$, subject to the constraint

$$\mathbf{J}_{\phi\phi} - \mathbf{J}_{\phi\tilde{\alpha}}\mathbf{J}_{\tilde{\alpha}\tilde{\alpha}}^{-1}\mathbf{J}_{\phi\tilde{\alpha}}^T \succeq \mathbf{U}. \tag{7.17}$$

Then, based on the Schur complement condition (Zhang, 2005), the constraint (7.16b) with a modified FIM can be obtained. □

Remark 7.1 Proposition 7.1 can be easily extended to the cases of multiple targets. For example, consider a system with Q targets. Denote α_q, $\phi_{q,\mathrm{h}}$, and $\phi_{q,\mathrm{v}}$ as the complex amplitude, the azimuth DoA, and the elevation DOA of the q-th target, respectively. The unknown parameters become $\bar{\boldsymbol{\xi}} = [\bar{\boldsymbol{\phi}}^T, \bar{\tilde{\boldsymbol{\alpha}}}^T]$, where $\bar{\boldsymbol{\phi}} = [\phi_{1,h}, \phi_{1,v}, \ldots, \phi_{Q,h}, \phi_{Q,v}]^T \in \mathbb{R}^{2Q\times 1}$ and $\bar{\tilde{\boldsymbol{\alpha}}} = [\mathrm{Re}(\alpha_1), \mathrm{Im}(\alpha_1), \ldots, \mathrm{Re}(\alpha_Q), \mathrm{Im}(\alpha_Q)] \in \mathbb{R}^{2Q\times 1}$. Following the same path in Appendix H, it can be shown that the corresponding CRB matrix of DOA estimation is in the same form as (7.15) but with a different dimension of $2Q \times 2Q$. Therefore, by defining a new auxiliary semidefinite matrix $\bar{\mathbf{U}} \in \mathbb{C}^{2Q\times 2Q}$, the corresponding CRB minimization problem for multiple targets can be converted into the same form as problem (7.16).

Based on Proposition 7.1, the optimization problem for minimizing CRB($\boldsymbol{\phi}$) can be formulated as follows:

$$\mathcal{P}: \quad \min_{\substack{\mathbf{U},\mathbf{P},\mathbf{R}_s,\\ \boldsymbol{\theta}_t,\boldsymbol{\theta}_r}} \quad \mathrm{tr}\left(\mathbf{U}^{-1}\right), \tag{7.18a}$$

$$\text{s.t.} \quad \begin{bmatrix} \mathbf{J}_{\phi\phi} - \mathbf{U} & \mathbf{J}_{\phi\tilde{\alpha}} \\ \mathbf{J}_{\phi\tilde{\alpha}}^T & \mathbf{J}_{\tilde{\alpha}\tilde{\alpha}} \end{bmatrix} \succeq 0, \tag{7.18b}$$

$$\gamma_k \geq \overline{\gamma}_k, \forall k \tag{7.18c}$$

$$\mathrm{tr}(\mathbf{P}\mathbf{P}^H + \mathbf{R}_s) \leq P, \tag{7.18d}$$

$$\beta_{t,n}^2 + \beta_{r,n}^2 = 1, 0 \leq \beta_{t,n}, \beta_{r,n} \leq 1, \forall n, \tag{7.18e}$$

$$\mathbf{R}_s \succeq 0, \mathbf{U} \succeq 0, \tag{7.18f}$$

where $\boldsymbol{\theta}_i = [\beta_{i,1}e^{j\varphi_{i,1}}, \ldots, \beta_{i,N}e^{j\varphi_{i,N}}]^T, \forall i \in \{t, r\}$, is the vector comprised of the entries on the diagonal of the matrix $\boldsymbol{\Theta}_i, \forall i \in \{t, r\}$, $\overline{\gamma}_k \geq 0$ in constraint (7.18c) denotes the minimum SINR threshold of user k, and $P \geq 0$ in constraint (7.18d) denotes the total power budget at the BS. Constraint (7.18e) characterizes the amplitude relationships between TCs and RCs of the STARS. Finally, constraint (7.18f) ensures the matrices $\mathbf{R}_s$ and $\mathbf{U}$ to be positive semidefinite. The main challenges for solving problem $\mathcal{P}$ are the highly coupled transmit waveform matrices, i.e., $\mathbf{P}$ and $\mathbf{R}_s$, and the STARS coefficient vectors, i.e., $\boldsymbol{\theta}_t$ and $\boldsymbol{\theta}_r$, in the non-convex constraints (7.18b) and (7.18c). As a consequence, it is challenging to find the global optimum of problem $\mathcal{P}$. Therefore, in the following subsections, we develop an efficient PDD-based algorithm to obtain a high-quality solution to problem $\mathcal{P}$.

7.3.2 PDD Framework for Solving Problem

As shown in Shi and Hong (2020), the standard PDD optimization framework is developed in a double-loop structure, where the AL problem of the original problem is optimized in a BCD manner in the inner loop, while the Lagrangian dual variables and penalty factors are updated in the outer loop. Therefore, the key idea of applying the PDD method for solving $\mathcal{P}$ is to construct an AL problem of it that has a simple or even closed-form solution at each step of the BCD. To this end, we first define auxiliary variables as follows:

$$\mathbf{F} = \mathbf{\Theta}_r \mathbf{G} \mathbf{R}_x \mathbf{G}^H \mathbf{\Theta}_r^H. \tag{7.19}$$

Then, problem $\mathcal{P}$ can be reformulated as

$$\min_{\chi} \quad \mathrm{tr}(\mathbf{U}^{-1}) \tag{7.20a}$$

$$\text{s.t.} \quad \mathbf{F} = \mathbf{\Theta}_r \mathbf{G} \mathbf{R}_x \mathbf{G}^H \mathbf{\Theta}_r^H, \tag{7.20b}$$

$$\begin{bmatrix} \mathbf{J}_{\boldsymbol{\phi}\boldsymbol{\phi}}(\mathbf{F}) - \mathbf{U} & \mathbf{J}_{\boldsymbol{\phi}\tilde{\boldsymbol{\alpha}}}(\mathbf{F}) \\ \mathbf{J}_{\boldsymbol{\phi}\tilde{\boldsymbol{\alpha}}}^T(\mathbf{F}) & \mathbf{J}_{\tilde{\boldsymbol{\alpha}}\tilde{\boldsymbol{\alpha}}}(\mathbf{F}) \end{bmatrix} \succeq 0, \tag{7.20c}$$

$$\text{(7.18c)–(7.18f)}, \tag{7.20d}$$

where $\boldsymbol{\chi} \triangleq \{\mathbf{U}, \mathbf{P}, \mathbf{R}_s, \mathbf{F}, \boldsymbol{\theta}_t, \boldsymbol{\theta}_r\}$ represents all optimization variables. The entries of the matrix in the left-hand side of constant (7.20c) are given by

$$\mathbf{J}_{\boldsymbol{\phi}\boldsymbol{\phi}}(\mathbf{F}) = \frac{2|\alpha|^2 L}{\sigma_s^2} \mathrm{Re}\left(\begin{bmatrix} \mathrm{tr}(\dot{\mathbf{B}}_{\phi_h} \mathbf{F} \dot{\mathbf{B}}_{\phi_h}^H) & \mathrm{tr}(\dot{\mathbf{B}}_{\phi_h} \mathbf{F} \dot{\mathbf{B}}_{\phi_v}^H) \\ \mathrm{tr}(\dot{\mathbf{B}}_{\phi_h} \mathbf{F} \dot{\mathbf{B}}_{\phi_v}^H) & \mathrm{tr}(\dot{\mathbf{B}}_{\phi_v} \mathbf{F} \dot{\mathbf{B}}_{\phi_v}^H) \end{bmatrix} \right), \tag{7.21}$$

$$\mathbf{J}_{\boldsymbol{\phi}\tilde{\boldsymbol{\alpha}}}(\mathbf{F}) = \frac{2L}{\sigma_s^2} \mathrm{Re}\left(\begin{bmatrix} \alpha^* \mathrm{tr}\left(\mathbf{B}\mathbf{F}\dot{\mathbf{B}}_{\phi_h}^H\right) \\ \alpha^* \mathrm{tr}\left(\mathbf{B}\mathbf{F}\dot{\mathbf{B}}_{\phi_v}^H\right) \end{bmatrix} [1, j] \right), \tag{7.22}$$

$$\mathbf{J}_{\tilde{\boldsymbol{\alpha}}\tilde{\boldsymbol{\alpha}}}(\mathbf{F}) = \frac{2L}{\sigma_s^2} \mathbf{I}_2 \mathrm{tr}\left(\mathbf{B}\mathbf{F}\mathbf{B}^H\right). \tag{7.23}$$

By introducing the Lagrangian dual variable $\mathbf{\Upsilon} \in \mathbb{C}^{N \times N}$ and the penalty factor $\rho > 0$ for the equality constraints (7.20b), the following AL problem of (7.20) can be obtained:

$$\mathcal{P}_{\mathrm{AL}}(\rho, \mathbf{\Upsilon}): \quad \min_{\chi} \quad \mathrm{tr}(\mathbf{U}^{-1}) + P_\rho(\boldsymbol{\chi}, \mathbf{\Upsilon}) \tag{7.24a}$$

$$\text{s.t.} \quad \text{(7.18c)–(7.18f)}, \ \text{(7.20c)}, \tag{7.24b}$$

Algorithm 8 PDD-based algorithm for solving problem $\mathcal{P}$

Initialize feasible $\boldsymbol{\chi}^{[0]}$, $\boldsymbol{\Upsilon}^{[0]}$, $\rho^{[0]} \geq 0$, and set $0 < c < 1, n = 1$.
repeat
  $\boldsymbol{\chi}^{[n+1]} = \text{optimize}\left(\mathcal{P}_{\text{AL}}(\rho^{[n]}, \boldsymbol{\Upsilon}^{[n]})\right)$.
  if $h(\boldsymbol{\chi}^{[n+1]}) \leq \eta^{[n]}$ **then**
    $\boldsymbol{\Upsilon}^{[n+1]} = \boldsymbol{\Upsilon}^{[n]} + \frac{1}{\rho}(\mathbf{F}^{[n+1]} - \boldsymbol{\Theta}_r^{[n+1]}\mathbf{G}\mathbf{R}_x^{[n+1]} \times \mathbf{G}^H(\boldsymbol{\Theta}_r^{[n+1]})^H)$.
    $\rho^{[n+1]} = \rho^{[n]}$.
  else
    $\boldsymbol{\Upsilon}^{[n+1]} = \boldsymbol{\Upsilon}^{[n]}$, $\rho^{[n+1]} = c\rho^{[n]}$.
  end if
  $n = n + 1$.
until the constraint violation $h(\boldsymbol{\chi})$ falls below a predefined threshold.

where

$$P_\rho(\boldsymbol{\chi}, \boldsymbol{\Upsilon}) = \frac{1}{2\rho}\|\mathbf{F} - \boldsymbol{\Theta}_r\mathbf{G}\mathbf{R}_x\mathbf{G}^H\boldsymbol{\Theta}_r^H + \rho\boldsymbol{\Upsilon}\|^2. \tag{7.25}$$

According to Shi and Hong (2020), the PDD-based algorithm for solving problem $\mathcal{P}$ is summarized in Algorithm 8, where $h(\boldsymbol{\chi})$ is the constraint violation function defined as follows:

$$h(\boldsymbol{\chi}) = \|\mathbf{F} - \boldsymbol{\Theta}_r\mathbf{G}\mathbf{R}_x\mathbf{G}^H\boldsymbol{\Theta}_r^H\|_\infty. \tag{7.26}$$

Moreover, $\{\eta^{[n]}\}_{n=1}^{\infty}$ is a sequence that converges to zero, which can be set empirically. In this chapter, we set $\eta^{[n]} = 0.99h(\boldsymbol{\chi}^{[n-1]})$. It can be observed that when the penalty factor ρ is sufficiently small, the penalty term $P_\rho(\boldsymbol{\chi})$ and the constraint violation $h(\boldsymbol{\chi})$ will reduce to zero. In other words, the equality constraint (7.20b) is satisfied. A detailed discussion of the convergence and optimality of the PDD framework can be found in Shi and Hong (2020).

7.3.3 Proposed BCD Algorithm for Solving AL Problem (7.24)

The key step of the proposed PDD-based algorithm is to solve the AL problem (7.24). In particular, we divide the set of the optimal variables $\boldsymbol{\chi}$ into two blocks, namely $\{\mathbf{U}, \mathbf{F}, \mathbf{P}, \mathbf{R}_s\}$ and $\{\boldsymbol{\theta}_t, \boldsymbol{\theta}_r\}$. Then, the BCD algorithm is invoked to solve each block iteratively while fixing the others, which leads to the following two subproblems.

7.3.3.1 Subproblem with Respect to $\{\mathbf{U}, \mathbf{F}, \mathbf{P}, \mathbf{R}_s\}$

The subproblem with respect to $\{\mathbf{U}, \mathbf{F}, \mathbf{P}, \mathbf{R}_s\}$ is given by

$$\min_{\mathbf{U},\mathbf{F},\mathbf{P},\mathbf{R}_s} \operatorname{tr}(\mathbf{U}^{-1}) + \frac{1}{2\rho}\|\mathbf{F} - \mathbf{\Theta}_r\mathbf{G}\mathbf{R}_x\mathbf{G}^H\mathbf{\Theta}_r^H + \rho\mathbf{\Upsilon}\|^2 \tag{7.27a}$$

$$\text{s.t.} \quad \begin{bmatrix} \mathbf{J}_{\phi\phi}(\mathbf{F}) - \mathbf{U} & \mathbf{J}_{\phi\tilde{\alpha}}(\mathbf{F}) \\ \mathbf{J}_{\phi\tilde{\alpha}}^T(\mathbf{F}) & \mathbf{J}_{\tilde{\alpha}\tilde{\alpha}}(\mathbf{F}) \end{bmatrix} \succeq 0, \tag{7.27b}$$

$$\frac{1}{\overline{\gamma}_k}|\mathbf{u}_k^H\mathbf{p}_k|^2 \geq \sum_{i\in\mathcal{K}\setminus k} |\mathbf{u}_k^H\mathbf{p}_i|^2 + \mathbf{u}_k^H\mathbf{R}_s\mathbf{u}_k + \sigma_k^2, \forall k, \tag{7.27c}$$

$$\operatorname{tr}(\mathbf{P}\mathbf{P}^H + \mathbf{R}_s) \leq P, \tag{7.27d}$$

$$\mathbf{R}_s \succeq 0, \mathbf{U} \succeq 0, \tag{7.27e}$$

where $\mathbf{u}_k = \mathbf{G}^H\mathbf{\Theta}_t^H\mathbf{h}_k$ denotes the effective channel vector for user k. Constraint (7.27c) is transformed from the communication SINR constraint (7.18c). Although problem (7.27) is apparently non-convex due to the constraint (7.27c), we now show that the global optimum of it can be obtained via the SDR approach.

Lemma 7.1 *Given* $\mathbf{P} = [\mathbf{p}_1, \ldots, \mathbf{p}_K]^T$, *there exists* $\mathbf{R}_x$ *and* $\mathbf{R}_s$ *that satisfy* $\mathbf{R}_x = \mathbf{P}\mathbf{P}^H + \mathbf{R}_s$ *if and only if*

$$\mathbf{R}_x \succeq \mathbf{P}\mathbf{P}^H = \sum_{k\in\mathcal{K}} \mathbf{p}_k\mathbf{p}_k^H. \tag{7.28}$$

Proof The necessity and sufficiency of this condition can be readily proved. □

By applying Lemma 7.1 and defining the auxiliary variables $\mathbf{P}_k = \mathbf{p}_k\mathbf{p}_k^H, \forall k \in \mathcal{K}$, that satisfy $\mathbf{P}_k \succeq 0$ and $\operatorname{rank}(\mathbf{P}_k) = 1$, we can formulate the SDR problem of (7.27) as follows:

$$\min_{\substack{\mathbf{U},\mathbf{F},\mathbf{R}_x,\\ \{\mathbf{P}_k\}_{k\in\mathcal{K}}}} \operatorname{tr}(\mathbf{U}^{-1}) + \frac{1}{2\rho}\|\mathbf{F} - \mathbf{\Theta}_r\mathbf{G}\mathbf{R}_x\mathbf{G}^H\mathbf{\Theta}_r^H + \rho\mathbf{\Upsilon}\|^2 \tag{7.29a}$$

$$\text{s.t.} \quad (1 + \frac{1}{\overline{\gamma}_k})\mathbf{u}_k^H\mathbf{P}_k\mathbf{u}_k \geq \mathbf{u}_k^H\mathbf{R}_x\mathbf{u}_k + \sigma_k^2, \forall k, \tag{7.29b}$$

$$\operatorname{tr}(\mathbf{R}_x) \leq P, \tag{7.29c}$$

$$\mathbf{R}_x \succeq \sum_{k\in\mathcal{K}} \mathbf{P}_k, \mathbf{R}_s \succeq 0, \mathbf{U} \succeq 0, \mathbf{P}_k \succeq 0, \forall k, \tag{7.29d}$$

where the non-convex rank-one constraints $\operatorname{rank}(\mathbf{P}_k) = 1, \forall k \in \mathcal{K}$, are relaxed. The relaxed problem (7.29) is convex SDP, the global optimum of which can be efficiently obtained by the existing convex optimization solvers. However, due to the omission of the rank-one constraints, the global optimum of problem (7.29) may have a higher rank, which may not be a solution to the original non-convex problem (7.27). Fortunately, in the following proposition, we show that the rank-one global optimum of problem (7.27) can always be constructed from an arbitrary global optimum of problem (7.29).

Proposition 7.2 *Given an arbitrary global optimum* $\tilde{\mathbf{R}}_x, \{\tilde{\mathbf{P}}_k\}_{k\in\mathcal{K}}$ *of problem* (7.29)*, the following solution is a global optimum of problem* (7.27)*:*

$$\mathbf{R}_x^\star = \tilde{\mathbf{R}}_x, \quad \mathbf{p}_k^\star = (\mathbf{u}_k^H \tilde{\mathbf{P}}_k \mathbf{u}_k)^{-1/2} \tilde{\mathbf{P}}_k \mathbf{u}_k. \tag{7.30}$$

Proof Please refer to Liu et al. (2020, Theorem 1) □

According to Proposition 7.2, a global optimum of problem (7.27) can be obtained by solving problem (7.29). Then, the optimal $\mathbf{R}_s$ can be calculated as

$$\mathbf{R}_s^\star = \mathbf{R}_x^\star - \sum_{k\in\mathcal{K}} \mathbf{p}_k^\star (\mathbf{p}_k^\star)^H. \tag{7.31}$$

7.3.3.2 Subproblem with Respect to $\{\boldsymbol{\theta}_t, \boldsymbol{\theta}_r\}$

The subproblem with respect to $\{\boldsymbol{\theta}_t, \boldsymbol{\theta}_r\}$ is given by

$$\min_{\boldsymbol{\theta}_t,\boldsymbol{\theta}_r} \quad \|\mathbf{F} - \boldsymbol{\Theta}_r \mathbf{G} \mathbf{R}_x \mathbf{G}^H \boldsymbol{\Theta}_r^H + \rho \boldsymbol{\Upsilon}\|^2 \tag{7.32a}$$

$$\text{s.t.} \quad \gamma_k \geq \overline{\gamma}_k, \forall k, \tag{7.32b}$$

$$\beta_{t,n}^2 + \beta_{r,n}^2 \leq 1, 0 \leq \beta_{t,n}, \beta_{r,n} \leq 1, \forall n. \tag{7.32c}$$

To facilitate the optimization of $\boldsymbol{\theta}_t$ and $\boldsymbol{\theta}_r$, we first transform the objective function and the communication SINR γ_k into the more trackable forms. In particular, we define $\bar{\mathbf{R}}_x = \mathbf{G}\mathbf{R}_x\mathbf{G}^H$. The eigenvalue decomposition of the matrix $\bar{\mathbf{R}}_x$ is given by

$$\bar{\mathbf{R}}_x = \sum_{j=1}^{R} \varrho_j \mathbf{v}_j \mathbf{v}_j^H = \sum_{j=1}^{R} \bar{\mathbf{v}}_j \bar{\mathbf{v}}_j^H, \tag{7.33}$$

where $\bar{\mathbf{v}}_j = \sqrt{\varrho_j}\mathbf{v}_j$ with ϱ_j and $\mathbf{v}_j$ denoting the eigenvalue and the corresponding eigenvector, respectively, and R denotes the rank of the matrix $\bar{\mathbf{R}}_x$. As such, the objective function can be reformulated as

$$\|\mathbf{F} - \sum_{j=1}^{R} \boldsymbol{\Theta}_r \bar{\mathbf{v}}_j \bar{\mathbf{v}}_j^H \boldsymbol{\Theta}_r^H + \rho\boldsymbol{\Upsilon}\|^2 = \|\bar{\mathbf{F}} - \sum_{j=1}^{R} \text{diag}(\bar{\mathbf{v}}_j)\boldsymbol{\theta}_r\boldsymbol{\theta}_r^H \text{diag}(\bar{\mathbf{v}}_j)^H + \rho\boldsymbol{\Upsilon}\|^2. \tag{7.34}$$

Next, we define the following variables:

$$\boldsymbol{\Phi}_{k,i} = \text{diag}(\mathbf{h}_k^H)\mathbf{G}\mathbf{p}_i\mathbf{p}_i^H\mathbf{G}^H\text{diag}(\mathbf{h}_k), \tag{7.35}$$

$$\boldsymbol{\Psi}_k = \text{diag}(\mathbf{h}_k^H)\mathbf{G}\mathbf{R}_s\mathbf{G}^H\text{diag}(\mathbf{h}_k), \tag{7.36}$$

$$\bar{\mathbf{V}}_j = \text{diag}(\bar{\mathbf{v}}_j), \bar{\mathbf{F}} = \mathbf{F} + \rho\boldsymbol{\Upsilon}. \tag{7.37}$$

As a consequence, problem (7.32) can be rewritten as

$$\min_{\boldsymbol{\theta}_t,\boldsymbol{\theta}_r} \quad \|\bar{\mathbf{F}} - \sum_{j=1}^{R} \bar{\mathbf{V}}_j\boldsymbol{\theta}_r\boldsymbol{\theta}_r^H\bar{\mathbf{V}}_j^H\|^2 \tag{7.38a}$$

$$\text{s.t.} \quad \frac{1}{\bar{\gamma}_k}\boldsymbol{\theta}_t^H\boldsymbol{\Phi}_{k,k}^*\boldsymbol{\theta}_t \geq \sum_{i\in\mathcal{K}\setminus k}\boldsymbol{\theta}_t^H\boldsymbol{\Phi}_{k,i}^*\boldsymbol{\theta}_t + \boldsymbol{\theta}_t^H\boldsymbol{\Psi}_k^*\boldsymbol{\theta}_t + \sigma_k^2, \forall k, \tag{7.38b}$$

$$|[\boldsymbol{\theta}_t]_n|^2 + |[\boldsymbol{\theta}_r]_n|^2 = 1, \forall n \in \mathcal{N}. \tag{7.38c}$$

By observing that the objective function and all the constraints of problem (7.38) are in quadratic form with respect to the optimization variables, we also exploit the SDR approach to approximately solve it. By defining the auxiliary variables $\mathbf{Q}_i = \boldsymbol{\theta}_i\boldsymbol{\theta}_i^H, \forall i \in \{t, r\}$, which satisfies $\mathbf{Q}_i \succeq 0$ and $\text{rank}(\mathbf{Q}_i) = 1$, the SDR problem of problem (7.38) is given by

$$\min_{\mathbf{Q}_t,\mathbf{Q}_r} \quad \|\bar{\mathbf{F}} - \sum_{k=1}^{R} \tilde{\mathbf{V}}_k\mathbf{Q}_r\tilde{\mathbf{V}}_k^H\|^2 \tag{7.39a}$$

$$\text{s.t.} \quad \frac{1}{\bar{\gamma}_k}\text{tr}(\boldsymbol{\Phi}_{k,k}^*\mathbf{Q}_t) \geq \sum_{i\in\mathcal{K}\setminus k}\text{tr}(\boldsymbol{\Phi}_{k,i}^*\mathbf{Q}_t) + \text{tr}(\boldsymbol{\Psi}_k^*\mathbf{Q}_t) + \sigma_k^2, \forall k, \tag{7.39b}$$

$$[\mathbf{Q}_t]_{n,n} + [\mathbf{Q}_r]_{n,n} = 1, \forall n \in \mathcal{N}, \tag{7.39c}$$

$$\mathbf{Q}_t \succeq 0, \mathbf{Q}_r \succeq 0. \tag{7.39d}$$

The above problem is also a convex SDP, and thereby the global optimum can be efficiently obtained via the existing convex optimization solvers. While the SDR may result in the solution with the general rank, the eigenvalue decomposition or Gaussian randomization (Luo et al., 2010) can be applied to construct a feasible rank-one solution to problem (7.38). Note that a sufficient number of Gaussian randomizations can achieve at least $\frac{\pi}{4}$-approximation of the optimal objective value of problem (7.38). However, the slight performance loss caused by constructing the rank-one solution cannot theoretically guarantee the monotonicity of the objective value during the BCD iteration process. In this case, the penalty-based alternating minimization (AltMin) algorithm proposed in Yu et al. (2021) with the provable convergence to the stationary point can be exploited, where the rank-one constraints

Algorithm 9 BCD algorithm for solving problem (7.24)

1: Initialize feasible χ.
2: **repeat**
3: Update $\{\mathbf{U}, \mathbf{F}, \mathbf{P}, \mathbf{R}_s\}$ by solving problem (7.29).
4: Update $\{\boldsymbol{\theta}_t, \boldsymbol{\theta}_r\}$ by solving problem (7.39).
5: **until** the fractional reduction of the objective value falls below a predefined threshold.

are transformed as a penalty term in the objective function. Nevertheless, it is worth mentioning that the eigenvalue decomposition or Gaussian randomization can generally guarantee the convergence of the proposed algorithm in practice.

The overall BCD algorithm for solving problem (7.24) is summarized in Algorithm 9. Typically, when the AltMin algorithm is exploited for updating the block $\{\boldsymbol{\theta}_t, \boldsymbol{\theta}_r\}$, Algorithm 9 is guaranteed to converge to a stationary point of problem (7.24) in polynomial time (Razaviyayn et al., 2013). The main complexity of Algorithm 9 arises from solving problems (7.29) and (7.39). Given the solution accuracy ϵ, the corresponding complexity via the interior-point method is in order of $\mathcal{O}((K^{6.5}M^{6.5} + N^{6.5})\log(1/\epsilon))$ and $\mathcal{O}((K + N)^{6.5}N^{6.5}\log(1/\epsilon))$, respectively (Liu et al., 2020; Toh, 2008).

7.4 STARS-ISAC: CRB Optimization with the Coupled Phase-Shift Model

In this section, we turn our attention to CRB optimization in the case of STARS with coupled phase shifts. Following a similar path in Sect. 7.3, the PDD framework is also invoked. Regarding the coupled phase-shift constraint, a low-complexity iterative algorithm is proposed, where the amplitude and phase-shift coefficients of SATRS are updated alternately by the closed-form solutions.

7.4.1 Problem Formulation

According to Proposition 7.1, the optimization problem for minimizing CRB($\boldsymbol{\phi}$) with the coupled phase-shift constraints of STARS can be formulated as follows:

$$\tilde{\mathcal{P}}: \quad \min_{\mathbf{U},\mathbf{P},\mathbf{R}_s,\boldsymbol{\theta}_t,\boldsymbol{\theta}_r} \quad \mathrm{tr}\left(\mathbf{U}^{-1}\right), \tag{7.40a}$$

$$\text{s.t.} \quad \cos(\varphi_{t,n} - \varphi_{r,n}) = 0, \forall n, \tag{7.40b}$$

$$(7.18\text{b})\text{–}(7.18\text{f}), \tag{7.40c}$$

where the coupled phase-shift constraint (7.40b) can be reformulated as follows:

$$|\varphi_{t,n} - \varphi_{r,n}| = \frac{1}{2}\pi \text{ or } \frac{3}{2}\pi, \forall n. \tag{7.41}$$

Compared to the independent phase-shift model, the coupled phase-shift model imposed in (7.40b) makes the problem even more complex. In particular, the optimization subject to this constraint requires hybrid continuous and discrete control. For example, the phase shift $\varphi_{t,n}$ can be selected as any value in a continuous region $[0, 2\pi]$, while the phase shift can only be selected from a discrete set $\{\varphi_{t,n} \pm \frac{1}{2}\pi, \varphi_{t,n} \pm \frac{3}{2}\pi\}$, which cannot be solved by existing methods. As a consequence, in the following subsections, we develop a new efficient PDD-based algorithm to solve the coupled phase-shift constraints.

7.4.2 PDD Framework for Solving Problem

In Sect. 7.3, we have proposed a PDD-based algorithm for the independent phase-shift model. In this case, we aim to obtain an equivalent form of problem (7.40) where the coupled phased shifts are relaxed to the independent ones. Toward this idea, in addition to $\mathbf{F} = \mathbf{\Theta}_r \mathbf{G}\mathbf{R}_x \mathbf{G}^H \mathbf{\Theta}_r^H$, we define another set of auxiliary variables as follows:

$$\tilde{\boldsymbol{\theta}}_i = \boldsymbol{\theta}_i, \forall i \in \{t, r\}, \tag{7.42}$$

where $\tilde{\boldsymbol{\theta}}_i = [\tilde{\beta}_{i,1} e^{j\tilde{\varphi}_{i,1}}, \ldots, \tilde{\beta}_{i,N} e^{j\tilde{\varphi}_{i,N}}]^T, \forall i \in \{t, r\}$. Then, problem $\tilde{\mathcal{P}}$ can be reformulated as

$$\min_{\tilde{\chi}} \quad \text{tr}(\mathbf{U}^{-1}) \tag{7.43a}$$

$$\text{s.t.} \quad \tilde{\boldsymbol{\theta}}_r = \boldsymbol{\theta}_r, \tilde{\boldsymbol{\theta}}_t = \boldsymbol{\theta}_t, \tag{7.43b}$$

$$\tilde{\beta}_{t,n}^2 + \tilde{\beta}_{r,n}^2 = 1, 0 \leq \tilde{\beta}_{t,n}, \tilde{\beta}_{r,n} \leq 1, \forall n, \tag{7.43c}$$

$$|\tilde{\varphi}_{t,n} - \tilde{\varphi}_{r,n}| = \frac{1}{2}\pi \text{ or } \frac{3}{2}\pi, \forall n, \tag{7.43d}$$

$$\beta_{t,n}^2 + \beta_{r,n}^2 = 1, 0 \leq \beta_{t,n}, \beta_{r,n} \leq 1, \forall n, \tag{7.43e}$$

$$\text{(7.18c)–(7.18f), (7.20b), (7.20c),} \tag{7.43f}$$

where $\tilde{\chi} \triangleq \{\mathbf{U}, \mathbf{P}, \mathbf{R}_s, \mathbf{F}, \boldsymbol{\theta}_t, \boldsymbol{\theta}_r, \tilde{\boldsymbol{\theta}}_t, \tilde{\boldsymbol{\theta}}_r\}$ represents all optimization variables. Note that in problem (7.43), the phase-shift constraints of the original optimization variables $\boldsymbol{\theta}_t$ and $\boldsymbol{\theta}_r$ have been relaxed to be independent, and the coupled phase-shift constraints are only related to the auxiliary variables $\tilde{\boldsymbol{\theta}}_t$ and $\tilde{\boldsymbol{\theta}}_r$. By introducing the Lagrangian dual variable $\boldsymbol{\lambda}_i \in \mathbb{C}^{N \times 1}, \forall i \in \{t, r\}$, for the additional equality constraints (7.43b), the following AL problem of (7.43) can be obtained:

Algorithm 10 PDD-based algorithm for solving problem $\tilde{\mathcal{P}}$

Initialize feasible $\tilde{\chi}^{[0]}$, $\boldsymbol{\Upsilon}^{[0]}$, $\boldsymbol{\lambda}_i^{[0]}, \forall i \in \{t, r\}$, $\rho^{[0]} \geq 0$, and set $0 < c < 1, n = 1$.
repeat
    $\tilde{\chi}^{[n+1]} = \text{optimize}\left(\tilde{\mathcal{P}}_{\text{AL}}(\rho^{[n]}, \boldsymbol{\Upsilon}^{[n]}, \boldsymbol{\lambda}_i^{[n]})\right)$
    if $h(\tilde{\chi}^{[n+1]}) \leq \eta^{[n]}$ **then**
        $\boldsymbol{\Upsilon}^{[n+1]} = \boldsymbol{\Upsilon}^{[n]} + \frac{1}{\rho}(\mathbf{F}^{[n+1]} - \boldsymbol{\Theta}_r^{[n+1]}\mathbf{G}\mathbf{R}_x^{[n+1]} \times \mathbf{G}^H(\boldsymbol{\Theta}_r^{[n+1]})^H)$.
        $\boldsymbol{\lambda}_i^{[n+1]} = \boldsymbol{\lambda}_i^{[n]} + \frac{1}{\rho}(\tilde{\boldsymbol{\theta}}_i^{[n+1]} - \boldsymbol{\theta}_i^{[n+1]}), \forall i \in \{t, r\}$.
        $\rho^{[n+1]} = \rho^{[n]}$.
    else
        $\boldsymbol{\Upsilon}^{[n+1]} = \boldsymbol{\Upsilon}^{[n]}, \boldsymbol{\lambda}_i^{[n+1]} = \boldsymbol{\lambda}_i^{[n]}, \forall i \in \{t, r\}$.
        $\rho^{[n+1]} = c\rho^{[n]}$.
    end if
    $n = n + 1$.
until the constraint violation $\tilde{h}(\tilde{\chi})$ falls below a predefined threshold.

$$\tilde{\mathcal{P}}_{\text{AL}}(\rho, \boldsymbol{\Upsilon}, \boldsymbol{\lambda}_i): \quad \min_{\tilde{\chi}} \quad \text{tr}(\mathbf{U}^{-1}) + \tilde{P}_\rho(\tilde{\chi}, \boldsymbol{\Upsilon}, \boldsymbol{\lambda}_i) \tag{7.44a}$$

$$\text{s.t.} \quad (7.18\text{c})\text{–}(7.18\text{f}),\ (7.20\text{b}),\ (7.20\text{c}),\ (7.43\text{b})\text{–}(7.43\text{e}), \tag{7.44b}$$

where

$$\tilde{P}_\rho(\tilde{\chi}, \boldsymbol{\Upsilon}, \boldsymbol{\lambda}_i) = \frac{1}{2\rho}\left(\|\mathbf{F} - \boldsymbol{\Theta}_r\mathbf{G}\mathbf{R}_x\mathbf{G}^H\boldsymbol{\Theta}_r^H + \rho\boldsymbol{\Upsilon}\|^2 + \sum_{i \in \{t,r\}} \|\tilde{\boldsymbol{\theta}}_i - \boldsymbol{\theta}_i + \rho\boldsymbol{\lambda}_i\|^2\right). \tag{7.45}$$

Thus, the PDD-based algorithm for solving problem $\tilde{\mathcal{P}}$ is summarized in Algorithm 10, and the constraint violation function is defined as

$$\tilde{h}(\tilde{\chi}) = \max\left\{\|\mathbf{F} - \boldsymbol{\Theta}_r\mathbf{G}\mathbf{R}_x\mathbf{G}^H\boldsymbol{\Theta}_r^H\|_\infty, \|\tilde{\boldsymbol{\theta}}_t - \boldsymbol{\theta}_t\|_\infty, \|\tilde{\boldsymbol{\theta}}_r - \boldsymbol{\theta}_r\|_\infty\right\}. \tag{7.46}$$

7.4.3 Proposed BCD Algorithm for Solving AL Problem (7.44)

Similar to Sect. 7.3.3, the BCD is exploited to solve the AL problem (7.44), where the set of optimization variables $\tilde{\chi}$ is divided into three blocks, namely $\{\mathbf{U}, \mathbf{F}, \mathbf{P}, \mathbf{R}_s\}$, $\{\boldsymbol{\theta}_t, \boldsymbol{\theta}_r\}$, and $\{\tilde{\boldsymbol{\theta}}_t, \tilde{\boldsymbol{\theta}}_r\}$. The solutions of the corresponding subproblems are given as follows.

7.4.3.1 Subproblem with Respect to $\{\mathbf{U}, \mathbf{F}, \mathbf{P}, \mathbf{R}_s\}$

The subproblem with respect to $\{\mathbf{U}, \mathbf{F}, \mathbf{P}, \mathbf{R}_s\}$ for the coupled phase-shift model is the same as that for the independent phase-shift model. Thus, it can be solved following the same path in Sect. 7.3.3.1.

7.4.3.2 Subproblem with Respect to $\{\boldsymbol{\theta}_t, \boldsymbol{\theta}_r\}$

The subproblem with respect to $\{\boldsymbol{\theta}_t, \boldsymbol{\theta}_r\}$ can be optimized with the relaxed independent phase-shit constraints, which is given by

$$\min_{\boldsymbol{\theta}_r, \boldsymbol{\theta}_t} \quad \|\mathbf{F} - \boldsymbol{\Theta}_r \mathbf{G} \mathbf{R}_x \mathbf{G}^H \boldsymbol{\Theta}_r^H + \rho \boldsymbol{\Upsilon}\|^2 + \sum_{i \in \{t,r\}} \|\tilde{\boldsymbol{\theta}}_i - \boldsymbol{\theta}_i + \rho \boldsymbol{\lambda}_i\|^2 \tag{7.47a}$$

$$\text{s.t.} \quad \gamma_k \geq \overline{\gamma}_k, \forall k \tag{7.47b}$$

$$\beta_{t,n}^2 + \beta_{r,n}^2 = 1, 0 \leq \beta_{t,n}, \beta_{r,n} \leq 1, \forall n. \tag{7.47c}$$

Then, by defining $\tilde{\boldsymbol{v}}_i = -(\tilde{\boldsymbol{\theta}}_i + \rho \boldsymbol{\lambda}_i), \forall i \in \{t, r\}$, and following a similar path in Sect. 7.3.3, the objective function can be reformulated as

$$\begin{aligned} &\|\bar{\mathbf{F}} - \sum_{j=1}^{R} \bar{\mathbf{V}}_j \boldsymbol{\theta}_r \boldsymbol{\theta}_r^H \bar{\mathbf{V}}_j^H \|^2 + \sum_{i \in \{t,r\}} \|\tilde{\boldsymbol{v}}_i + \boldsymbol{\theta}_i\|^2 \\ =&\|\bar{\mathbf{F}} - \sum_{j=1}^{R} \hat{\mathbf{V}}_j \boldsymbol{\vartheta}_r \boldsymbol{\vartheta}_r^H \hat{\mathbf{V}}_j^H \|^2 + \sum_{i \in \{t,r\}} \left(\boldsymbol{\vartheta}_i^H \boldsymbol{\Xi}_i \boldsymbol{\vartheta}_i + \tilde{\boldsymbol{v}}_i^H \tilde{\boldsymbol{v}}_i \right), \end{aligned} \tag{7.48}$$

where

$$\hat{\mathbf{V}}_j = \left[\bar{\mathbf{V}}_j \ \mathbf{0}_N \right], \boldsymbol{\Xi}_i = \begin{bmatrix} \mathbf{I}_N & \tilde{\boldsymbol{v}}_i \\ \tilde{\boldsymbol{v}}_i^H & 0 \end{bmatrix}, \boldsymbol{\vartheta}_i = \begin{bmatrix} \kappa_i \boldsymbol{\theta}_i \\ \kappa_i \end{bmatrix}, \tag{7.49}$$

and $|\kappa_i|^2 = 1$. Similarly, the minimum communication SINR can be reformulated as

$$\frac{1}{\overline{\gamma}_k} \boldsymbol{\vartheta}_t^H \hat{\boldsymbol{\Phi}}_{k,k}^* \boldsymbol{\vartheta}_t \geq \sum_{i \in \mathcal{K} \setminus k} \boldsymbol{\vartheta}_t^H \hat{\boldsymbol{\Phi}}_{k,i}^* \boldsymbol{\vartheta}_t + \boldsymbol{\vartheta}_t^H \hat{\boldsymbol{\Psi}}_k^* \boldsymbol{\vartheta}_t + \sigma_k^2, \forall k, \tag{7.50}$$

where

$$\hat{\boldsymbol{\Phi}}_{k,i} = \begin{bmatrix} \boldsymbol{\Phi}_{k,i} & \mathbf{0}_N \\ \mathbf{0}_N^T & 0 \end{bmatrix}, \quad \hat{\boldsymbol{\Psi}}_k = \begin{bmatrix} \boldsymbol{\Psi}_k & \mathbf{0}_N \\ \mathbf{0}_N^T & 0 \end{bmatrix}. \tag{7.51}$$

It can be observed that the objective function and all the constraints of problem (7.47) have been transformed into the homogeneous quadratic form. Thus, we also exploit the SDR approach to approximately solve it. In particular, by defining the auxiliary variables $\hat{\mathbf{Q}}_i = \boldsymbol{\vartheta}_i \boldsymbol{\vartheta}_i^H, \forall i \in \{t, r\}$, which satisfies $\hat{\mathbf{Q}}_i \succeq 0$ and $\text{rank}(\hat{\mathbf{Q}}_i) = 1$, the SDR problem of (7.47) is given by

$$\min_{\hat{\mathbf{Q}}_t, \hat{\mathbf{Q}}_r} \quad \|\bar{\mathbf{F}} - \sum_{j=1}^{R} \hat{\mathbf{V}}_j \hat{\mathbf{Q}}_t \hat{\mathbf{V}}_j^H\|^2 + \sum_{i \in \{t,r\}} \text{tr}(\boldsymbol{\Xi}_i \hat{\mathbf{Q}}_i) \tag{7.52a}$$

$$\text{s.t.} \quad \frac{1}{\overline{\gamma}_k} \text{tr}(\hat{\boldsymbol{\Phi}}_{k,k}^* \hat{\mathbf{Q}}_t) \geq \sum_{i \in \mathcal{K} \setminus k} \text{tr}(\hat{\boldsymbol{\Phi}}_{k,i}^* \hat{\mathbf{Q}}_t) + \text{tr}(\hat{\boldsymbol{\Psi}}_k^* \hat{\mathbf{Q}}_t) + \sigma_k^2, \forall k, \tag{7.52b}$$

$$[\hat{\mathbf{Q}}_t]_{n,n} + [\hat{\mathbf{Q}}_r]_{n,n} = 1, \forall n, \tag{7.52c}$$

$$[\hat{\mathbf{Q}}_t]_{N+1,N+1} = [\hat{\mathbf{Q}}_r]_{N+1,N+1} = 1, \tag{7.52d}$$

$$\hat{\mathbf{Q}}_t \succeq 0, \hat{\mathbf{Q}}_r \succeq 0, \tag{7.52e}$$

which is a convex SDP. Denote $\boldsymbol{\vartheta}_i^\star, \forall i \in \{t, r\}$, as the approximated rank-one solution. Then, the corresponding solution of problem (7.47) is given by

$$\boldsymbol{\theta}_i^\star = \frac{1}{|[\boldsymbol{\vartheta}_i^\star]_{N+1}|} [\boldsymbol{\vartheta}_i^\star]_{1:N}, \forall i \in \{t, r\}. \tag{7.53}$$

Similarly, the penalty-based AltMin method in Yu et al. (2021) can be employed to guarantee the theoretical convergence of the BCD.

7.4.3.3 Subproblem with Respect to $\{\tilde{\boldsymbol{\theta}}_t, \tilde{\boldsymbol{\theta}}_r\}$

The optimization variables $\{\tilde{\boldsymbol{\theta}}_t, \tilde{\boldsymbol{\theta}}_r\}$ only appear in the penalty term in the objective function and the coupled amplitude and phase-shift constraints. Thus, the subproblem with respect to $\{\tilde{\boldsymbol{\theta}}_t, \tilde{\boldsymbol{\theta}}_r\}$ is given by

$$\min_{\tilde{\boldsymbol{\theta}}_t, \tilde{\boldsymbol{\theta}}_r} \quad \sum_{i \in \{t,r\}} \|\tilde{\boldsymbol{\theta}}_i - \boldsymbol{\theta}_i + \rho \boldsymbol{\lambda}_i\|^2 \tag{7.54a}$$

$$\text{s.t.} \quad \tilde{\beta}_{t,n}^2 + \tilde{\beta}_{r,n}^2 = 1, 0 \leq \tilde{\beta}_{t,n}, \tilde{\beta}_{r,n} \leq 1, \forall n, \tag{7.54b}$$

$$|\tilde{\varphi}_{t,n} - \tilde{\varphi}_{r,n}| = \frac{1}{2}\pi \text{ or } \frac{3}{2}\pi, \forall n. \tag{7.54c}$$

In this problem, both constraints are non-convex, and the second constraint even requires a binary decision, which is challenging to solve. However, in the following, we show that the amplitude coefficients and the phase-shift coefficients can be optimized alternately using the closed-form solutions. To this end, we reformulate

$\tilde{\boldsymbol{\theta}}_i$ as

$$\tilde{\boldsymbol{\theta}}_i = \text{diag}(\tilde{\boldsymbol{\beta}}_i)\tilde{\mathbf{q}}_i, \forall i \in \{t, r\}, \tag{7.55}$$

where $\tilde{\boldsymbol{\beta}}_i = [\tilde{\beta}_{i,1}, \ldots, \tilde{\beta}_{i,N}]^T$ and $\tilde{\mathbf{q}}_i = [e^{j\tilde{\varphi}_{i,1}}, \ldots, e^{j\tilde{\varphi}_{i,N}}]^T$. By defining $\boldsymbol{v}_i = -\boldsymbol{\theta}_i + \rho\boldsymbol{\lambda}_i, \forall i \in \{t, r\}$, the objective function of (7.54) can be reformulated into

$$\begin{aligned}
&\sum_{i\in\{t,r\}} \|\tilde{\boldsymbol{\theta}}_i - \boldsymbol{\theta}_i + \rho\boldsymbol{\lambda}_i\|^2 = \sum_{i\in\{t,r\}} \|\text{diag}(\tilde{\boldsymbol{\beta}}_i)\tilde{\mathbf{q}}_i + \boldsymbol{v}_i\|^2 \\
&= \sum_{i\in\{t,r\}} 2\text{Re}(\boldsymbol{v}_i^H \text{diag}(\tilde{\boldsymbol{\beta}}_i)\tilde{\mathbf{q}}_i) + \sum_{i\in\{t,r\}} \tilde{\mathbf{q}}_i^H \text{diag}(\tilde{\boldsymbol{\beta}}_i)\text{diag}(\tilde{\boldsymbol{\beta}}_i)\tilde{\mathbf{q}}_i + \sum_{i\in\{t,r\}} \boldsymbol{v}_i^H \boldsymbol{v}_i \\
&\overset{(a)}{=} \sum_{i\in\{t,r\}} 2\text{Re}(\boldsymbol{v}_i^H \text{diag}(\tilde{\boldsymbol{\beta}}_i)\tilde{\mathbf{q}}_i) + \sum_{i\in\{t,r\}} \sum_{n\in\mathcal{N}} \tilde{\beta}_{i,n}^2 + \sum_{i\in\{t,r\}} \boldsymbol{v}_i^H \boldsymbol{v}_i \\
&\overset{(b)}{=} \sum_{i\in\{t,r\}} 2\text{Re}(\boldsymbol{v}_i^H \text{diag}(\tilde{\boldsymbol{\beta}}_i)\tilde{\mathbf{q}}_i) + \underbrace{N + \sum_{i\in\{t,r\}} \boldsymbol{v}_i^H \boldsymbol{v}_i}_{\text{constant}},
\end{aligned} \tag{7.56}$$

where the equality (a) is due to the property $|[\tilde{\mathbf{q}}_i]_n|^2 = |e^{j\tilde{\varphi}_{i,n}}|^2 = 1$, and the equality (b) stems from the property $\tilde{\beta}_{t,n}^2 + \tilde{\beta}_{r,n}^2 = 1$. By removing the constant term in the objective function, problem (7.54) can be simplified as

$$\min_{\tilde{\boldsymbol{\beta}}_t, \tilde{\boldsymbol{\beta}}_r, \tilde{\mathbf{q}}_t, \tilde{\mathbf{q}}_r} \quad \sum_{i\in\{t,r\}} \text{Re}(\boldsymbol{v}_i^H \text{diag}(\tilde{\boldsymbol{\beta}}_i)\tilde{\mathbf{q}}_i) \tag{7.57a}$$

$$\text{s.t.} \quad (7.54\text{b}), (7.54\text{c}). \tag{7.57b}$$

To solve it, we first give the following two propositions.

Proposition 7.3 (Closed-Form Solution for Coupled Phase Shifts) *With the coupled phase-shift constraint* (7.54c), *for any given* $\tilde{\boldsymbol{\beta}}_t$ *and* $\tilde{\boldsymbol{\beta}}_r$, *the optimal solution for the n-th entries* $\tilde{q}_{t,n}$, $\tilde{q}_{r,n}$ *of* $\tilde{\mathbf{q}}_t$, $\tilde{\mathbf{q}}_r$ *is chosen from the following two pairs of solutions*

$$\begin{cases} \tilde{q}_{t,n} = e^{j(\pi - \angle\psi_n^+)}, \tilde{q}_{r,n} = e^{j(\frac{3}{2}\pi - \angle\psi_n^+)}, \\ \tilde{q}_{t,n} = e^{j(\pi - \angle\psi_n^-)}, \tilde{q}_{r,n} = e^{j(\frac{1}{2}\pi - \angle\psi_n^-)}, \end{cases} \tag{7.58}$$

where $\psi_n^+ = \tilde{v}_{t,n}^* + j\tilde{v}_{r,n}^*$ *and* $\psi_n^- = \tilde{v}_{t,n}^* - j\tilde{v}_{r,n}^*$, *such that the value of* $\text{Re}(\tilde{v}_{t,n}^*\tilde{q}_{t,n}) + \text{Re}(\tilde{v}_{r,n}^*\tilde{q}_{r,n})$ *is minimized. Here,* $\tilde{v}_{i,n}^*$ *denotes the n-th entry of the vector* $\tilde{\boldsymbol{v}}_i^H = \boldsymbol{v}_i^H \text{diag}(\tilde{\boldsymbol{\beta}}_i)$, $\forall i \in \{t, r\}$.

Proof Please refer to Appendix I. □

Algorithm 11 AO algorithm for solving problem (7.54)

1: Initialize feasible $\tilde{\mathbf{q}}_t$, $\tilde{\mathbf{q}}_r$, $\tilde{\boldsymbol{\beta}}_t$, and $\tilde{\boldsymbol{\beta}}_r$.
2: **repeat**
3: Update each entry of $\tilde{\mathbf{q}}_t$ and $\tilde{\mathbf{q}}_r$ by (7.58).
4: Update each entry of $\tilde{\boldsymbol{\beta}}_t$ and $\tilde{\boldsymbol{\beta}}_r$ by (7.59).
5: **until** the fractional reduction of the objective value falls below a predefined threshold.

Algorithm 12 BCD algorithm for solving problem (7.44)

1: Initialize feasible $\tilde{\chi}$ such that $\mathbf{F} = \boldsymbol{\Theta}_r \mathbf{G} \mathbf{R}_x \mathbf{G}^H \boldsymbol{\Theta}_r^H$ and $\tilde{\boldsymbol{\theta}}_i = \boldsymbol{\theta}_i, \forall i \in \{t, r\}$.
2: **repeat**
3: Update $\{\mathbf{U}, \mathbf{F}, \mathbf{P}, \mathbf{R}_s\}$ by solving problem (7.29).
4: Update $\{\boldsymbol{\theta}_t, \boldsymbol{\theta}_r\}$ by solving problem (7.52).
5: Update $\{\tilde{\boldsymbol{\theta}}_t, \tilde{\boldsymbol{\theta}}_r\}$ by solving problem (7.54) through Algorithm 11.
6: **until** the fractional reduction of the objective value falls below a predefined threshold.

Proposition 7.4 (Closed-Form Solution for Amplitudes) *For any given $\tilde{\mathbf{q}}_t$ and $\tilde{\mathbf{q}}_r$, the optimal solution of the n-th entries $\tilde{\beta}_{t,n}$, $\tilde{\beta}_{r,n}$ of $\tilde{\boldsymbol{\beta}}_t$, $\tilde{\boldsymbol{\beta}}_r$ is given by*

$$\tilde{\beta}_{t,n} = \sin \omega_n, \tilde{\beta}_{r,n} = \cos \omega_n, \tag{7.59}$$

$$\omega_n = \begin{cases} -\frac{1}{2}\pi - \psi_n, & \textit{if } \psi_n \in [-\pi, -\frac{1}{2}\pi), \\ 0, & \textit{if } \psi_n \in [-\frac{1}{2}\pi, \frac{1}{4}\pi), \\ \frac{1}{2}\pi, & \textit{otherwise}, \end{cases} \tag{7.60}$$

where $\psi_n = \mathrm{sgn}(b_n) \arccos(\frac{a_n}{\sqrt{a_n^2+b_n^2}}) \in [-\pi, \pi]$, $a_n = |\breve{\upsilon}_{t,n}^*| \cos(\angle \breve{\upsilon}_{t,n}^*)$, $b_n = |\breve{\upsilon}_{r,n}^*| \cos(\angle \breve{\upsilon}_{r,n}^*)$, *and* $\breve{\upsilon}_{i,n}^*$ *is the n-th entry of the vector* $\breve{\boldsymbol{\upsilon}}_i^H = \boldsymbol{\upsilon}_i^H \mathrm{diag}(\tilde{\mathbf{q}}_i)$, $\forall i \in \{t, r\}$.

Proof Please refer to Appendix J. □

According to Proposition 7.3 and Proposition 7.4, by fixing the amplitude (phase-shift) coefficients, the phase-shift (amplitude) coefficients have the optimal closed-form solutions. Consequently, problem (7.54) can also be solved iteratively, the detail of which is given in Algorithm 11. Since the optimal solution is obtained at each step, the convergence of Algorithm 11 to the stationary points is guaranteed (Razaviyayn et al., 2013).

The overall BCD algorithm for solving problem (7.44) is summarized in Algorithm 12. Similarly, when the AltMin method is adopted for optimizing the block $\{\boldsymbol{\theta}_t, \boldsymbol{\theta}_r\}$, the convergence of Algorithm 12 to the stationary point can be theoretically guaranteed (Razaviyayn et al., 2013). The complexity of this algorithm is summarized as follows. Firstly, the complexities of using the interior-point method to solve problem (7.29) and (7.52) are in order of $O((K^{6.5}M^{6.5} + N^{6.5}) \log(1/\epsilon))$ and $O((K+N+2)^{6.5}(N+1)^{6.5} \log(1/\epsilon))$, respectively (Liu et al., 2020; Toh, 2008). Moreover, in Algorithm 11, the complexities of updating $\{\tilde{\mathbf{q}}_t, \tilde{\mathbf{q}}_r\}$ and $\{\tilde{\boldsymbol{\beta}}_t, \tilde{\boldsymbol{\beta}}_r\}$ are in order of $O(4N)$ and $O(2N)$, respectively.

7.5 Numerical Results

In this section, numerical results obtained through Monte Carlo simulations are provided to evaluate the performance of the proposed STARS-enabled ISAC system. We assume that the BS is 40 m away from the STARS. The communication users are randomly distributed within $20 \sim 50$ m from the transmission side of STARS, and the sensing target is 30 m away from the reflection side of the STARS. We assume the Rician channel model for all communication channels. Thus, the channels $\mathbf{G}$ and $\mathbf{h}_k, \forall k \in \mathcal{K}$, are given by

$$\mathbf{G} = \sqrt{\frac{\rho_0}{d_{BR}^{\alpha_{BR}}}} \left(\sqrt{\frac{\varepsilon}{1+\varepsilon}} \mathbf{G}^{\mathrm{LoS}} + \sqrt{\frac{1}{1+\varepsilon}} \mathbf{G}^{\mathrm{NLoS}} \right), \tag{7.61}$$

$$\mathbf{h}_k = \sqrt{\frac{\rho_0}{d_{RU,k}^{\alpha_{RU}}}} \left(\sqrt{\frac{\varepsilon}{1+\varepsilon}} \mathbf{h}_k^{\mathrm{LoS}} + \sqrt{\frac{1}{1+\varepsilon}} \mathbf{h}_k^{\mathrm{NLoS}} \right), \tag{7.62}$$

where d_{BR} and $d_{RU,k}$ denote the BS-STARS distance and the STARS-user-k distance, respectively, α_{BR} and α_{RU} denote the corresponding path loss exponents, ρ_0 denotes the path loss at the reference distance of 1 m, ε denotes the Rician factor, $\mathbf{G}^{\mathrm{LoS}}$ and $\mathbf{h}_k^{\mathrm{LoS}}$ are the deterministic LoS component, and $\mathbf{G}^{\mathrm{NLoS}}$ and $\mathbf{h}_k^{\mathrm{NLoS}}$ are the random non-LoS component modeled as Rayleigh fading. The main adopted system parameters are given in Table 7.1. The CVX toolbox (Grant & Boyd, 2014) is used to solve all convex problems involved in the proposed algorithms. The convergence threshold of the PDD-based algorithms and the BCD-based algorithms are set as 10^{-4} and 10^{-3}, respectively. Without loss of generality, we assume that all communication users have the same SINR requirement, i.e., $\overline{\gamma}_k = \overline{\gamma}, \forall k \in \mathcal{K}$.

To verify the efficiency of the proposed framework, we compared it with a baseline that employs one conventional reflecting-only RIS and one conventional

Table 7.1 System parameters

Parameter	Description	Value
M	The number of antennas at the BS	10
N	The number of passive elements at the STARS	6~18
N_s	The number of sensor elements	2~24
P	Transmit power at the BS	30 dBm
K	The number of communication users	4 or 6
σ_k^2, σ_s^2	The noise power	−110 dBm
α_{BR}, α_{RU}	Path loss exponents	2
L	The length of a coherent time block	100
ρ_0	The path loss at 1 m	30 dB
$\phi_{\mathrm{h}}, \phi_{\mathrm{v}}$	Azimuth and elevation DOAs of the target	120°, 30°
ε	Rician factor	3 dB

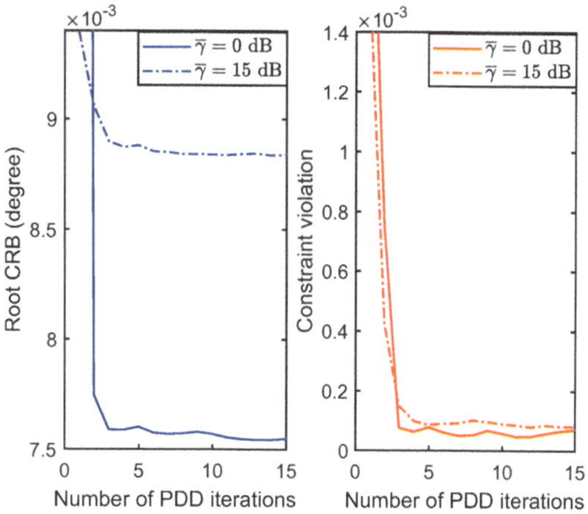

Fig. 7.4 The convergence behavior of Algorithm 8

transmitting-only RIS, both with $N/2$ elements and adjacent to each other at the same location as the STARS. This baseline is essentially a special case of STARS where the amplitudes of the TCs and RCs are fixed to $\boldsymbol{\beta}_t = [\mathbf{1}_{1\times N/2}, \mathbf{0}_{1\times N/2}]^T$ and $\boldsymbol{\beta}_r = [\mathbf{0}_{1\times N/2}, \mathbf{1}_{1\times N/2}]^T$. Therefore, the resultant optimization problem can also be solved by the proposed PDD-based algorithm. All the following numerical results are obtained by averaging over 50 random channel realizations unless otherwise specified.

7.5.1 Convergence Performance of the Proposed Algorithms

In Figs. 7.4 and 7.5, we examine the convergence performance of the proposed algorithms over a random channel realization. In particular, the initialization point of Algorithm 8 is randomly selected, while that of Algorithm 10 is selected as the output of Algorithm 8. Note that the original unit of the root CRB is radians, which is difficult to intuitively match the accuracy of actual DOA estimates. Therefore, we convert the unit of the root CRB from radians to degrees. In terms of the root CRB and the constraint violation, it can be observed that by using Algorithm 8 and Algorithm 10, they can converge well for both independent and coupled phase-shift models. Moreover, when the SINR threshold is low, namely $\overline{\gamma} = 0$dB, Algorithm 10 almost converges at the first PDD iteration. This is because, with the low SINR threshold, the coupled phase-shift model is almost identical to the independent phase-shift model. Consequently, when Algorithm 10 is initialized as the output of Algorithm 8, the optimal solution is almost reached at the beginning. This phenomenon will be further explained in the following numerical results.

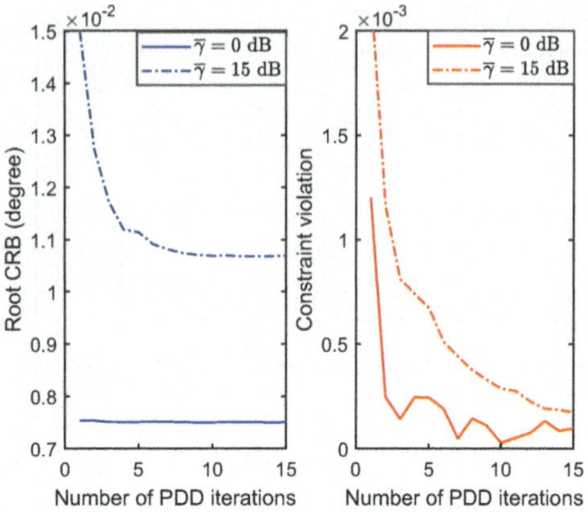

Fig. 7.5 The convergence behavior of Algorithm 10

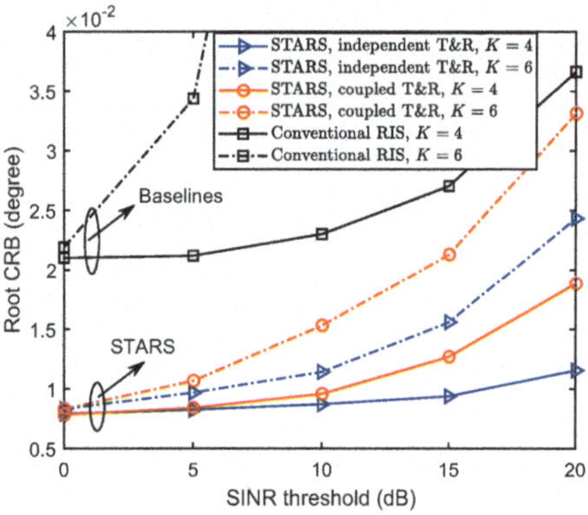

Fig. 7.6 Root CRB versus the communication SINR threshold $\overline{\gamma}$ for $N = 10$ and $N_s = 5$

7.5.2 Root CRB versus Communication SINR Threshold

In Fig. 7.6, we studied the achieved root CRB versus the communication SINR threshold considering different user numbers. As can be observed, there is a trade-off between the sensing and communication performance. This is because the higher communication performance requires more resources such as power and DoFs in the communication space, resulting in lower sensing performance. It can also be seen that independent of the phase-shift model, STARS always outperforms conventional RIS in achieving a lower CRB. The conventional RIS almost becomes infeasible when $K = 6$ and $\overline{\gamma} > 5$dB. This is indeed expected since conventional RIS utilizes a limited number of elements for sensing and communication spaces and therefore cannot achieve the same DoFs as STARS. Furthermore, the performance gap between coupled and independent phase shifts of the STARS is small when the communication SINR threshold $\overline{\gamma}$ is weak. However, as $\overline{\gamma}$ increases, the transmission phase shifts for the communication space become more stringent, which also limits the reflection phase shifts for the sensing space when the phase

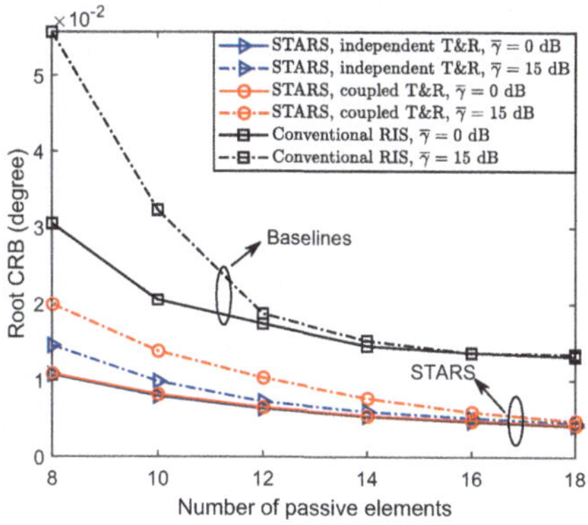

Fig. 7.7 Root CRB versus the number of passive elements N for $N_s = 5$ and $K = 4$

shifts are coupled, leading to a larger performance gap with the independent phase shifts.

7.5.3 Root CRB versus Number of Passive Elements

In Fig. 7.7, we illustrate the impact of the number of passive elements N of STARS when $N_s = 5$ and $K = 4$. It can be observed that as the number of passive elements increases, the root CRB decreases and the communication SINR threshold $\overline{\gamma}$ has less influence on the CRB performance. This is because more passive elements introduce more DoFs to construct a more directional sensing beam and achieve a more flexible communication channel tuning capability. Due to the same reason, when $\overline{\gamma} = 15$dB, the root CRB achieved by the two phase-shift models finally converges to the same value as N increases. However, when $\overline{\gamma} = 0$dB, the performance of the coupled phase-shift model is always comparable to that of the independent phase-shift model.

7.5.4 Root CRB versus Number of Sensor Elements

In Fig. 7.8, we further studied the impact of the number of active sensor elements N_s when $N = 10$. We can see that all schemes are capable of achieving higher estimation accuracy when more sensor elements are installed. Similarly, the superior performance gain of STARS compared with conventional RIS is also demonstrated. When $\overline{\gamma} = 0$dB, the coupled phase-shift model achieves almost the same performance as the independent phase-shift model. When $\overline{\gamma} = 15$dB, there is an obvious performance gap between the two phase-shift models, which is gradually reduced as the number of sensor elements increases.

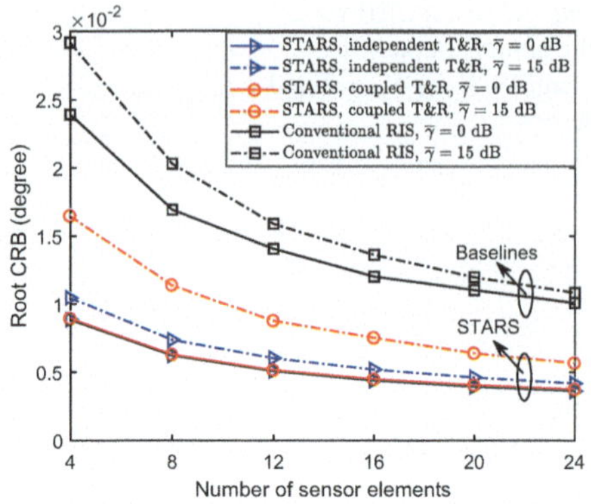

Fig. 7.8 Root CRB versus the number of sensor elements N_s for $N = 10$ and $K = 4$

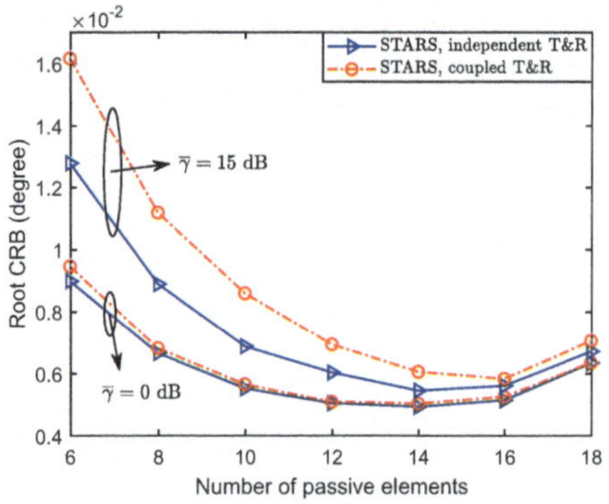

Fig. 7.9 Trade-off between the number of passive and sensor elements for $N + N_s = 20$ and $K = 4$

7.5.5 Trade-Off Between Passive and Sensor Elements

In Fig. 7.9, to obtain more insights, we plot the root CRB versus the number of passive elements when the total number of the passive and sensor elements is fixed to be $N + N_s = 20$. As can be observed, the more passive elements do not necessarily lead to lower root CRB in this case, which implies that there is a trade-off between the number of passive elements and the number of sensor elements. Furthermore, for both cases with $\overline{\gamma} = 0$ dB and $\overline{\gamma} = 15$ dB, the best sensing performance is achieved when $N = 14$ and $N_s = 6$. In other words, installing more passive elements can improve system performance more effectively than installing more sensor elements. This is because more passive elements provide more full-space DoFs for both C&S, while more sensor elements only increase the dimension of the signals used for sensing.

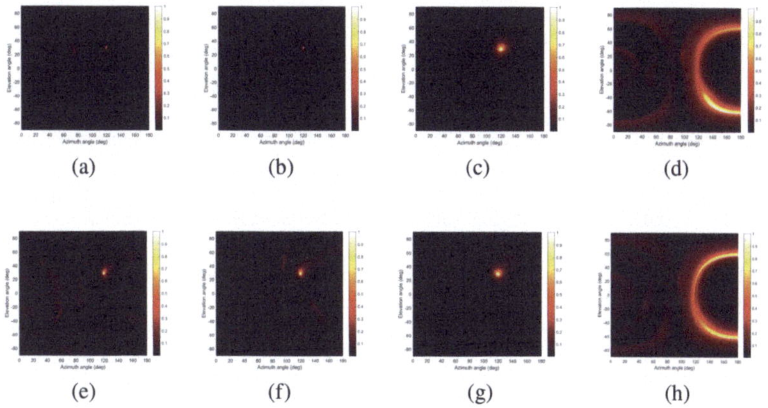

Fig. 7.10 Normalized spectrum of MLE by different schemes for $N = 10$, $N_s = 5$, and $K = 4$. (**a**) STARS, independent, $\overline{\gamma} = 0\,\text{dB}$. (**b**) STARS, coupled, $\overline{\gamma} = 0\,\text{dB}$. (**c**) Conventional RIS, $\overline{\gamma} = 0\,\text{dB}$. (**d**) STARS, random, $\overline{\gamma} = 0\,\text{dB}$. (**e**) STARS, independent, $\overline{\gamma} = 20\,\text{dB}$. (**f**) STARS, coupled, $\overline{\gamma} = 20\,\text{dB}$. (**g**) Conventional RIS, $\overline{\gamma} = 20\,\text{dB}$. (**h**) STARS, random, $\overline{\gamma} = 20\,\text{dB}$

7.5.6 Spectrum of MLE

To verify the effectiveness of optimizing the CRB, the practical MLE of DOAs ϕ_{h} and ϕ_{v} is investigated. Figure 7.10 demonstrates the normalized spectrum of MLE obtained via the 2D search over a predefined fine angle grid. In the spectrum, the point with the highest value (the brightest point) is corresponding to the estimated DOAs. As can be observed, when $\overline{\gamma} = 0\text{dB}$, both independent and coupled phase-shift STARSs can achieve a point-like brightest region around $(120°, 30°)$ in the spectrum, which indicates the high accuracy of estimating DOAs. When $\overline{\gamma}$ increases to 20 dB, the brightest region achieved by STARS is still relatively small. However, when the conventional RIS is exploited, the brightest region becomes very large regardless of the value of $\overline{\gamma}$, which implies a low sensing resolution. Finally, we also demonstrate the spectrum obtained by the STARS with random reflection phase shifts. In this case, we can see that the estimation of DOAs is almost infeasible, which verifies the effectiveness of optimizing the CRB.

7.6 Summary

This chapter discusses the employment of STARS for ISAC, where a novel sensing-at-STARS is introduced and numerical case studies are provided to demonstrate the effectiveness.

References

Bekkerman, I., & Tabrikian, J. (2006). Target detection and localization using MIMO radars and sonars. *IEEE Transactions on Signal Processing, 54*, 3873–3883.

Boyd, S., Boyd, S. P., & Vandenberghe, L. (2004). *Convex optimization*. Cambridge University Press.

Buzzi, S., Grossi, E., Lops, M., & Venturino, L. (2022). Foundations of MIMO radar detection aided by reconfigurable intelligent surfaces. *IEEE Transactions on Signal Processing, 70*, 1749–1763.

Grant, M., & Boyd, S. (2014). CVX: Matlab software for disciplined convex programming, version 2.1. http://cvxr.com/cvx

Kay, S. M. (1993). *Fundamentals of statistical signal processing: Estimation theory*. Prentice-Hall.

Liu, F., Cui, Y., Masouros, C., Xu, J., Han, T. X., Eldar, Y. C., & Buzzi, S. (2022). Integrated sensing and communications: Toward dual-functional wireless networks for 6G and beyond. *IEEE Journal on Selected Areas in Communications, 40*, 1728–1767.

Liu, F., Liu, Y.-F., Li, A., Masouros, C., & Eldar, Y. C. (2021). Cramér-Rao bound optimization for joint radar-communication beamforming. *IEEE Transactions on Signal Processing, 70*, 240–253.

Liu, X., Huang, T., Shlezinger, N., Liu, Y., Zhou, J., & Eldar, Y. C. (2020). Joint transmit beamforming for multiuser MIMO communications and MIMO radar. *IEEE Transactions on Signal Processing, 68*, 3929–3944.

Liu, Z., Li, X., Ji, H., Zhang, H., & Leung, V. C. M. (2023). Toward STAR-RIS-empowered integrated sensing and communications: Joint active and passive beamforming design. *IEEE Transactions on Vehicular Technology, 72*, 15991–16005.

Luo, Z.-Q., Ma, W.-K., So, A. M.-C., Ye, Y., & Zhang, S. (2010). Semidefinite relaxation of quadratic optimization problems. *IEEE Signal Processing Magazine, 27*, 20–34.

Manikas, A. (2004). *Differential geometry in array processing*. Imperial College Press.

Meng, K., Wu, Q., Chen, W., & Li, D. (2024). Sensing-assisted communication in vehicular networks with intelligent surface. *IEEE Transactions on Vehicular Technology, 73*, 876–893.

Razaviyayn, M., Hong, M., & Luo, Z.-Q. (2013). A unified convergence analysis of block successive minimization methods for nonsmooth optimization. *SIAM Journal on Optimization, 23*, 1126–1153.

Shao, X., You, C., Ma, W., Chen, X., & Zhang, R. (2022). Target sensing with intelligent reflecting surface: Architecture and performance. *IEEE Journal on Selected Areas in Communications, 40*, 2070–2084.

Shi, Q., & Hong, M. (2020). Penalty dual decomposition method for nonsmooth nonconvex optimization—Part I: Algorithms and convergence analysis. *IEEE Transactions on Signal Processing, 68*, 4108–4122.

Sun, W., Sun, S., Su, X., & Liu, R. (2023). Security-ensured integrated sensing and communication (ISAC) systems enabled by phase-coupled intelligent omni-surfaces (IOS). *IEEE Transactions on Wireless Communications*. https://doi.org/10.1109/TWC.2023.3308973

Toh, K.-C. (2008). An inexact primal–dual path following algorithm for convex quadratic SDP. *Programs in Mathematics, 112*, 221–254.

Wang, B., Li, H., Shen, S., Cheng, Z., & Clerckx, B. (2023a). A dual-function radar-communication system empowered by beyond diagonal reconfigurable intelligent surface. arXiv preprint arXiv:2301.03286.

Wang, C., Wang, C.-C., Li, Z., Ng, D. W. K., Wong, K.-K., Al-Dhahir, N., & Niyato, D. (2023b). STAR-RIS-enabled secure dual-functional radar-communications: Joint waveform and reflective beamforming optimization. *IEEE Transactions on Information Forensics and Security, 18*, 4577–4592.

Wang, Y., Yang, Z., Cui, J., Xu, P., Chen, G., Quek, T. Q. S., & Tafazolli, R. (2023). Optimizing the fairness of STAR-RIS and NOMA assisted integrated sensing and communication systems. *IEEE Transactions on Wireless Communications*. https://doi.org/10.1109/TWC.2023.3328872

Wu, C., You, C., Liu, Y., Gu, X., & Cai, Y. (2022). Channel estimation for STAR-RIS-aided wireless communication. *IEEE Communications Letters, 26*, 652–656.
Yu, X., Xu, D., Ng, D. W. K., & Schober, R. (2021). IRS-assisted green communication systems: Provable convergence and robust optimization. *IEEE Transactions on Communications, 69*, 6313–6329.
Zhang, F. (2005). *The Schur complement and its applications*. Springer.
Zhang, Z., Liu, Y., Wang, Z., and Chen, J. (2023). STARS-ISAC: How many sensors do we need? *IEEE Transactions on Wireless Communications, 23*(2), 1085–1099. https://doi.org/10.1109/TWC.2023.3285795

Part III
Challenges and Conclusions

Chapter 8
Research Directions and Conclusions

Abstract In this chapter, we will discuss several interesting research directions for STARS and conclude the book.

8.1 Research Directions for STARS

In this section, we discuss several interesting research directions for STARS toward 6G, namely NOMA communications, near-field communications (NFCs), PLS, simultaneous wireless information and power transfer (SWIPT), and mobile edge computing (MEC).

8.1.1 STARS-Aided NOMA

NOMA emerges as a significant contender for multiple access in future wireless networks, enabling multiple users to simultaneously share the same resource block across various domains such as time, frequency, and space (Liu et al., 2022). STARS can effectively and intelligently modify the wireless propagation environment, utilizing both transmission and reflection beamforming, to cater to the requirements of NOMA. In the following, we will provide an overview of pertinent research endeavors about STARS-aided downlink NOMA communications.

User clustering is critical to fully reap the benefits of NOMA in downlink systems to determine which users share the same resource blocks (Liu et al., 2022). Traditionally, user clustering is coordinated based on channel characteristics, including channel gain and channel correlation, to ensure that NOMA is effectively implemented within each clustering. Compared to conventional reflecting/transmitting-only RISs, STARSs provide additional flexibility for NOMA. On the one hand, in a STARS-NOMA system, user clustering can be carried out within each side of STARS, which is referred to as *unilateral user clustering (UUC)*, as shown in Fig. 8.1a. In UUC, the channel conditions of users in each cluster can only be

Y. Liu et al., *Simultaneously Transmitting and Reflecting Surfaces for Wireless Communications*, Wireless Networks, https://doi.org/10.1007/978-3-031-76848-4_8

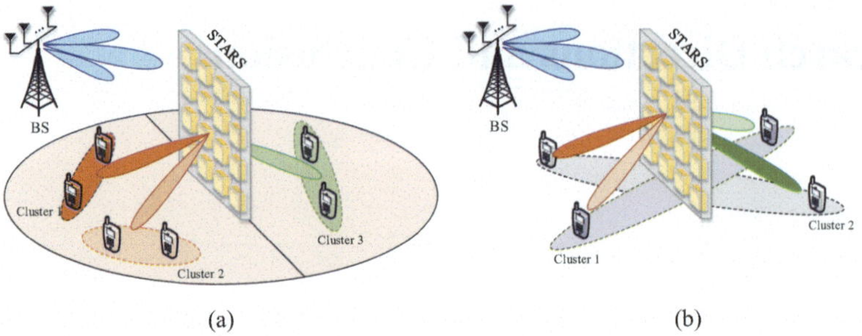

Fig. 8.1 Illustration of two user clustering strategies in STARS-aided downlink NOMA. (**a**) Unilateral user clustering. (**b**) Bilateral user clustering

adjusted by a single transmission or reflection beamforming matrix. On the other hand, as shown in Fig. 8.1b, STARSs also enable a new user clustering strategy that allows users on different sides to be allocated into one cluster, which is referred to *bilateral user clustering (BUC)*. In BUC, the channel conditions of users in each cluster can be configured by both the transmission and reflection beamforming matrices, thus providing more DoFs for facilitating NOMA. In the following, we review research works on the two user clustering strategies.

8.1.1.1 Unilateral User Clustering

Zuo et al. (2023) examined a STARS-NOMA system with the fixed UUC. A two-layer iterative algorithm was proposed for optimizing resource allocation, including decoding order, power allocation, BS beamforming, and STARS beamforming, to maximize the sum rate. Further studies in energy-efficient STARS-NOMA designs with UUC were provided in Fang et al. (2023) and Wang et al. (2023). The former introduced a heuristic zero-forcing method for joint optimization of BS and STARS beamforming matrices, while the latter proposed a convex optimization-based approach. To reduce system complexity, Asif et al. (2023) explored a hybrid NOMA strategy that dedicates individual time slots to each NOMA cluster, where an iterative algorithm was developed for joint optimization of decoding order, time allocation, and beamforming matrices. Lastly, Li et al. (2023b) investigated a STARS-assisted cognitive radio NOMA system, where one primary user and two clusters of secondary users are situated on different sides of the STARS. In Li et al. (2023b), the outage probability, ergodic rate, and EE for both primary and secondary users were analyzed.

8.1.1.2 Bilateral User Clustering

The simplest two-user case of BUC-based STARS-NOMA systems, involving one transmission user and one reflection user within one NOMA cluster, has garnered considerable research interest to obtain potential design insights (Yue et al., 2023; Yang et al., 2022; Xu et al., 2023b; Zhang et al., 2022a; Xu et al., 2023a; Wang et al., 2023). Specifically, Yue et al. (2023) divided STARS elements into two groups of equal size for transmission and reflection users and adopted a coherent PSC for STARS beamforming. Subsequently, the effect of the proposed approach on the outage probability and ergodic rate was analyzed considering both perfect and imperfect SIC cases. The ES protocol in two-user BUC-based STARS-NOMA systems was further investigated in Yang et al. (2022). To mitigate the impact of the energy-splitting ratio on NOMA decoding orders, Yang et al. (2022) proposed a QoS-based energy-splitting scheme and examined the associated outage probability. Advancing this research, Xu et al. (2023b) introduced a series of optimization-based power allocation and STARS beamforming methods for maximizing the ergodic rate. A comprehensive analysis and comparison of ES, MS, and TS protocols in STARS-NOMA were conducted in Zhang et al. (2022a), focusing on their impacts on the outage probabilities and diversity gains. Considering latency requirements, Xu et al. (2023a) explored the ergodic capacity and block error rate in two-user BUC-based STARS-NOMA systems. Additionally, Wang et al. (2023) conceived an FD transmission scheme for BUC-based STARS-NOMA. This approach aimed to enhance communication performance for two cell-edge users by employing a cell-center user as an FD decode-and-forward relay.

In STARS-NOMA systems with more than two users, the BUC approach becomes much more complex, as the effective channels of NOMA users are influenced by both transmission and reflection beamforming of STARS. A straightforward strategy to circumvent intricate user clustering in such scenarios is to group all transmission and reflection users into a single NOMA cluster. To this end, Zhu et al. (2023) proposed an index modulation scheme for STARS to support users in such a single NOMA cluster, where the STARS is divided into equally sized subsurfaces dedicated to specific transmission or reflection users. Expanding on this idea, Vu et al. (2023) adapted the index modulation scheme to short-packet transmission scenarios. Diverging from equal partitioning of STARS, Aldababsa et al. (2023) introduced a scalable two-stage method, aiming to find a near-optimal STARS element partitioning strategy to maximize achievable rates. To reduce the optimization and channel estimation complexity, Wu et al. (2023) proposed a pair of two-timescale transmission schemes for STARS-aided downlink NOMA systems, where the STARS beamforming and power allocation are optimized using the long-term and instantaneous channel information, respectively. Moreover, Zhang et al. (2023) studied the queue-aware STARS-aided NOMA system considering the stability issues. Apart from the aforementioned BUC approach with one cluster, the multi-cluster-based BUC approach has also been widely investigated. To address the resulting challenging user clustering issue, Wu et al. (2022) and Lei et al. (2024) proposed two matching theory-based BUC approaches by exploiting the effective channel gain and the possible maximum value of minimum user rate as utility functions, respectively. Abrar et al. (2023) introduced a low-complexity optimization-based BUC approach that transforms the complex user clustering problem into a simple linear programming problem, which achieves a near-optimal performance compared to the exhaustive search method. Wen et al. (2023) studied a BUC-based STARS-NOMA system with unbalanced user distributions on the transmission and reflection sides. To mitigate the unbalanced effect, a series of PSC and energy-splitting strategies were conceived in Wen et al. (2023), where the corresponding ergodic rates over Nakagami-*m* fading channels were analyzed.

8.1.2 STARS-Aided NFC

Both far-field and near-field effects are intrinsic EM phenomena associated with radio waves emitted by an antenna or an array of antennas. In the far-field region, wave propagation can typically be represented as planar waves, whereas in the near-field region, it has to be accurately modeled as spherical waves (Balanis, 2016). The simplicity of the far-field planar wave model has led to its widespread adoption in earlier generations of wireless networks. However, the emergence of large-scale antenna arrays, such as supermassive MIMO and CF mMIMO (Björnson et al., 2019), coupled with the utilization of high-frequency bands like mmWave and THz bands (Akyildiz et al., 2014), has made the near-field effect increasingly significant or even dominant in future wireless networks (Liu et al., 2023a). The extent of the near-field region is proportional to the antenna array's aperture and the signal frequency. For example, following classical Rayleigh distance criteria (Balanis, 2016), an antenna array with a 1-m aperture and a 28-GHz signal frequency could have a near-field region extending up to 370 m, making NFC design crucial for future wireless networks.

While the near-field effect leads to a more complex wave propagation model, it also facilitates a new form of beamforming design, i.e., *beamfocusing* (Zhang et al., 2022b; Mu et al., 2024). In contrast to far-field beamsteering, which only differentiates users in the angular domain, near-field beamfocusing can distinguish users in both angular and distance domains, thus greatly enhancing multiple access capabilities but at the cost of higher optimization and channel estimation overhead. Compared to conventional reflecting/transmitting-only RISs, STARSs with massive elements not only facilitate full-space near-field beamfocusing through the ES or TS protocol but also enable dynamic switching between far-field beamforming and near-field beamfocusing through the MS protocol. In the following, we will elaborate on the ES/TS-STARS-aided NFC and MS-STARS-aided NFC, respectively.

- **ES/TS-STARS-aided NFC:** Utilizing the ES/TS protocol can maximize the aperture size of STARS for transmission and/or reflection, thus creating a large near-field region on both sides, as shown in Fig. 8.2a. This setup enables effective near-field beamfocusing across the full space. However, ES/TS-STARS-aided NFC requires a massive number of STARS elements to be optimized for both transmission and reflection, as well as complicated channel estimation in both angular and distance domains. To reduce the optimization complexity, Li

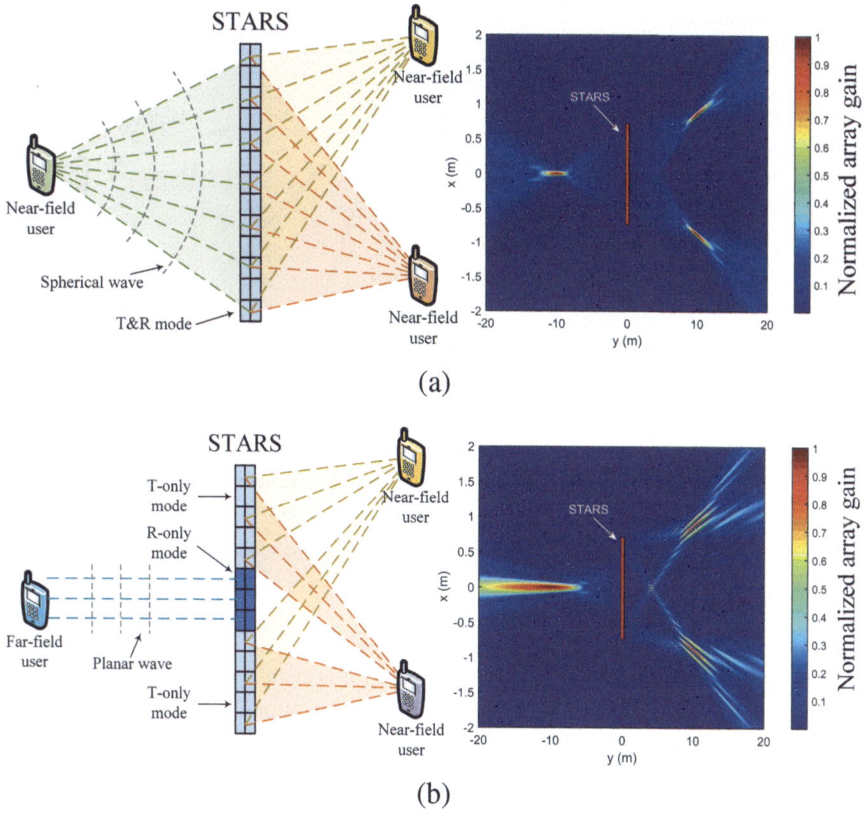

Fig. 8.2 Illustration of STARS-aided NFC and the beamforming patterns achieved using different operating protocols. For the generation of beamforming patterns, we consider a single-antenna BS and a 256-element STARS with 28 GHz carrier frequency and half-wavelength element spacing. (**a**) ES/TS-STARS (Near-field beamfocusing on both sides). (**b**) MS-STARS (Near-field beamfocusing and far-field beamsteering on two sides).

et al. (2023a) proposed a low-complexity element-wise STARS beamforming design. Furthermore, a unified near-far-field codebook-based STARS beamforming design was introduced in Zhang et al. (2024) to reduce the channel estimation overhead in STARS-NFC.

- **MS-STARS-aided NFC:** The MS protocol allows dynamic adjustment of the effective aperture size for transmission and reflection by selectively activating STARS elements in transmission or reflection mode (Xu et al., 2024), as shown in Fig. 8.2b. This enables switching transmission and reflection users between near-field and far-field regions. Therefore, different forms of beamforming can be realized on different sides of the STARS, thus satisfying specific requirements of transmission and reflection users, such as QoS and channel estimation overhead.

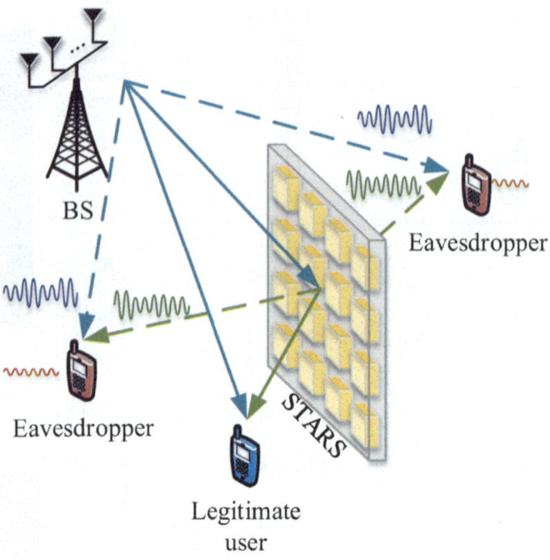

Fig. 8.3 Illustration of STARS-aided full-space PLS, where the received signal strength of eavesdroppers distributed on both sides can be mitigated via transmission and reflection

8.1.3 STARS-Aided PLS

Due to the broadcast nature of wireless channels, it is critical to prevent private data from being eavesdropped through wireless communications. To this end, PLS has been proposed as an efficient technology for achieving secrecy transmission (Zeng et al., 2019). Compared to conventional encryption methods and secret key management, the key idea of PLS is to exploit the characteristics of wireless channels (e.g., fading, interference, and noise) to combat malicious information eavesdropping. Therefore, with the capability of adjusting the signal propagation, RISs are beneficial for improving the performance of PLS. However, given the uncertain locations of malicious eavesdroppers, conventional reflecting-only RISs can only provide a half-space safeguarding wireless environment. As a remedy, a full-space safeguarding wireless environment can be facilitated by STARS, and secrecy transmission can be achieved regardless of eavesdroppers' locations, as shown in Fig. 8.3. Note that the extended full-space PLS also introduces complicated secrecy communication design problems, which has led to many research contributions on STARS-aided PLS in recent years. For example, Niu et al. (2021) first studied the joint beamforming design for a STARS-aided secure communication network, where the weighted sum secrecy rate was maximized under the three STARS operating protocols of STARS. Considering the imperfect CSI of both legitimate and eavesdropping channels, Wang et al. (2022) further investigated robust joint beamforming design to maximize the secrecy EE of a STARS-aided NOMA network. They showed that TS and ES/MS protocols are preferable for limited and high-power regimes, respectively. A similar STARS-aided secure NOMA communication design was also studied in Jia et al. (2023), where the worst secrecy rate among NOMA users was maximized given the imperfect CSI of eavesdroppers.

The STARS-aided PLS was studied by Zhang et al. (2023c) with the coupled phase-shift model, where the fair secrecy rate among users was maximized with the developed penalty-based optimization algorithm. Moreover, the study of STARS-aided PLS was extended to many advanced scenarios. Zhou et al. (2023) studied the employment of STARS for simultaneously anti-jamming and anti-eavesdropping, where ES was shown to be the optimal protocol to achieve the minimum power consumption for satisfying the secrecy communication requirement. Instead of using STARS for combating eavesdroppers, Hu et al. (2023) exploited the STARS for proactive eavesdropping of malicious transmission. In particular, STARS is deployed surrounding a legitimate monitor to not only assist in the jamming of malicious transmission but also eavesdrop the malicious information. Xie et al. (2023) further studied the STARS-aided PLS in a multicell NOMA network using the tool of stochastic geometry. In Xie et al. (2023), the secrecy outage probability and average secrecy capacity were derived, which showed that the ES protocol can achieve a better secrecy performance than the TS protocol.

8.1.4 STARS-Aided SWIPT

Sustainability plays a crucial role in the advancement of Internet of Things (IoT) and wireless sensor networks, laying the groundwork for numerous promising applications in the future, including smart homes, cities, factories, and environmental monitoring. SWIPT is widely recognized as a key enabler in extending the lifespan and mitigating the energy challenges faced by IoT and wireless sensor networks, especially in remote and inaccessible areas (Ponnimbaduge Perera et al., 2018). Note that employing the STARS in SWIPT can compensate for the disadvantage of energy leaking in conventional STARS-aided communication systems. For demonstration, let us take a STARS-aided unicast communication system as an example. Due to the STARS feature, the energy of signals intended for users on one side of the STARS will generally leak to the other side as harmful interference. This not only calls for interference management but also reduces EE. Fortunately, in STARS-aided SWIPT, the aforementioned leaked energy can be redesigned to improve the energy-harvesting performance of energy-harvesting users, as shown in Fig. 8.4. This, however, leads to a complicated STARS beamforming design for striking a good trade-off between information and energy-harvesting performance. To address this issue, Zhu et al. (2023) studied robust resource allocation for STARS-aided SWIPT under the imperfect CSI condition. Dual-objective resource allocation optimization problems for ES-STARS and TS-STARS were investigated in Zhu et al. (2023) to simultaneously maximize the energy-harvesting fairness and communication rate fairness, which characterized the fundamental rate-energy trade-off. It unveiled that ES-STARS is the optimal protocol for information-priority and energy-harvesting-priority cases, while TS-STARS is preferable to be employed for achieving a better rate-energy trade-off. Moreover, Yaswanth et al. (2023) studied the joint beamforming design for STARS-aided SWIPT for

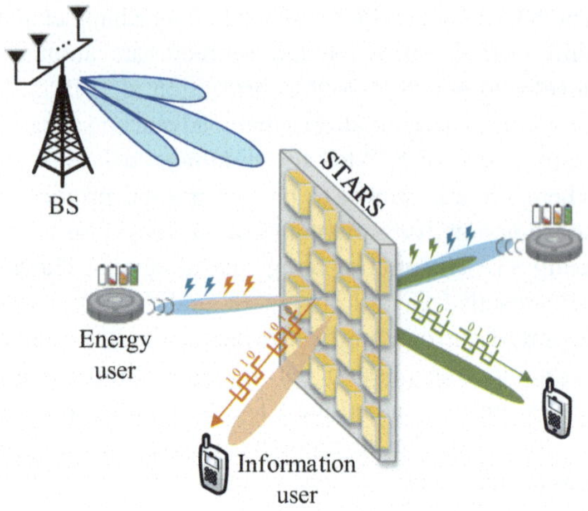

Fig. 8.4 Illustration of STARS-aided SWIPT

minimizing the transmit power subject to the minimum information and energy-harvesting requirements. Focusing on a basic STARS-aided SWIPT system with a single information/energy-harvesting user, Zhang et al. (2023a) conducted the performance analysis of the rate outage probability and energy outage probability at the high SNR regime, which also demonstrated the significant performance gain of STARS for SWIPT. Furthermore, STARS were deployed in wireless-powered communication systems in Du et al. (2023), Jiang et al. (2023). In Du et al. (2023), a TS-STARS was deployed in a wireless-powered IoT network to successively assist the wireless power charging phase from a power station to IoT devices and the wireless information uploading phase from IoT devices to an AP. Jiang et al. (2023) further exploit active STARS in a wireless-powered communication system, which consists of one downlink SWIPT stage and one uplink wireless information uploading stage. The results showed that active STARS can greatly improve the uplink sum rate compared to passive STARS.

8.1.5 STARS-Aided MEC

MEC enables mobile users to off-load complicated computational tasks to edge servers, thereby reducing energy consumption and computational latency (Mao et al., 2017). RISs have been recognized for their effectiveness in enhancing the off-loading link and minimizing off-loading delays (Bai et al., 2020; Zhou et al., 2021). However, conventional RISs, which only offer the reflecting or transmitting capability, are limited to supporting users in a half-space for task off-loading to the edge server. Furthermore, in RIS-MEC systems, users are constrained to off-load tasks only to the edge servers located on the same side as the RIS. These restrictions inherent in conventional RISs can lead to suboptimal utilization of computational

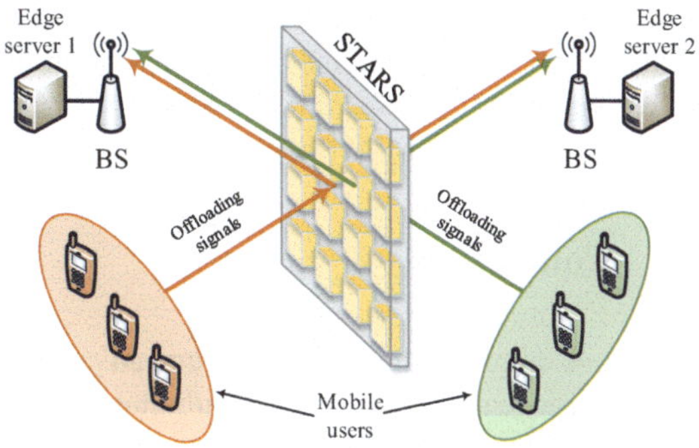

Fig. 8.5 Illustration of STARS-aided full-space MEC

resources and decreased overall computational efficiency. As a remedy, STARSs offer the capability for full-space MEC. This advancement enables edge servers to provide computational services to mobile users located on both sides of STARS, as shown in Fig. 8.5. Moreover, it allows mobile users to off-load tasks to servers situated on either side of STARS, thereby overcoming the spatial constraint inherent in traditional RIS-aided MEC. In Zhang et al. (2023b), the EE gain facilitated by TS-STARS in MEC systems was examined. The study introduced an AO-based approach for STARS beamforming design and time allocation to minimize the total energy consumption of mobile users. Liu et al. (2023b) explored MS-STARS in MEC systems, where a penalty-based iterative algorithm was proposed for optimizing STARS beamforming to maximize the overall computation rate. To extend the lifespan of mobile devices, Qin et al. (2023) integrated wireless power transfer into STARS-aided MEC. This approach initially charges transmission and reflection mobile users via wireless signals from the BS, who then off-loaded tasks using the charged energy. In a separate investigation, Zhai et al. (2023) studied a STARS-assisted over-the-air computation system, where the STARS beamforming under the coupled phase-shift model was optimized to minimize the computation mean-squared error. A similar STARS-aided MEC setup was further studied in Al-Habob et al. (2023), where STARS beamforming was optimized to reduce the latency for a group of mobile users off-loading tasks to multiple servers located on both sides of STARS.

8.1.6 Risk Mitigation for Malicious STARS

The passive, low-cost, and low-energy consumption features of STARS also introduce potential risks of malicious use. For instance, STARS can be employed to disrupt legitimate channels through interference (Rivetti et al., 2024) or to replicate legitimate channels for eavesdropping (Wei et al., 2023). Unlike conventional jamming schemes, the passive operation of STARS makes them difficult to detect. The situation could become more problematic with the deployment of large numbers of malicious STARS. Therefore, addressing the potential security issues posed by untrusted STARS is crucial before we can fully benefit from their advantages, necessitating further research efforts.

8.2 Conclusion

This book aims to discuss the recent advanced metasurface techniques, named STARS, for realizing sustainable, ubiquitous, and green future wireless communications. Specifically, the basic concept of STARS was introduced with different categories and hardware implementations. Moreover, the performance analysis and beamforming design for STARS were comprehensively discussed, which revealed the benefits of STARS for wireless communications. Furthermore, the employment of ML tools in STARS-aided wireless communications and STARS-aided ISAC designs were introduced. Finally, other interesting research directions with SATRS have been identified. Through all the provided case studies and numerical results, STARSs are shown to be an effective technology for 6G and beyond with facilitated full-space coverage.

References

Abrar, M. F. U., Talha, M., Ansari, R. I., Hassan, S. A., & Jung, H. (2023). Optimization of STAR-RIS-assisted hybrid NOMA mmWave communication. *IEEE Transactions on Vehicular Technology, 72*, 10146–10161.

Akyildiz, I. F., Jornet, J. M., & Han, C. (2014). Terahertz band: Next frontier for wireless communications. *Journal of Physics Communications. 12*, 16–32.

Al-Habob, A. A., Waqar, O., & Tabassum, H. (2023). Latency minimization in phase-coupled STAR-RIS assisted multi-MEC server systems. In *Proceedings of the IEEE International Symposium on Personal, Indoor and Mobile Radio Communications (PIMRC)* (pp. 1–7).

Aldababsa, M., Khaleel, A., & Basar, E. (2023). Simultaneous transmitting and reflecting reconfigurable intelligent surfaces-empowered NOMA networks. *IEEE Systems Journal, 17*, 5441–5451.

Asif, M., Ihsan, A., Khan, W. U., Ali, Z., Zhang, S., & Wu, S. X. (2023). Energy-efficient beamforming and resource optimization for STAR-IRS enabled hybrid-NOMA 6G communications. *IEEE Transactions on Green Communications and Networking, 7*, 1356–1368.

Bai, T., Pan, C., Deng, Y., Elkashlan, M., Nallanathan, A., & Hanzo, L. (2020). Latency minimization for intelligent reflecting surface aided mobile edge computing. *IEEE Journal on Selected Areas in Communications, 38*, 2666–2682.

Balanis, C. A. (2016). *Antenna theory: Analysis and design*. Wiley.

Björnson, E., Sanguinetti, L., Wymeersch, H., Hoydis, J., & Marzetta, T. L. (2019). Massive MIMO is a reality—what is next? Five promising research directions for antenna arrays. *Digital Signal Processing, 94*, 3–20.

Du, W., Chu, Z., Chen, G., Xiao, P., Xiao, Y., Wu, X., & Hao, W. (2023). STAR-RIS assisted wireless powered IoT networks. *IEEE Transactions on Vehicular Technology, 72*, 10644–10658.

Fang, F., Wu, B., Fu, S., Ding, Z., & Wang, X. (2023). Energy-efficient design of STAR-RIS aided MIMO-NOMA networks. *IEEE Transactions on Communications, 71*, 498–511.

Hu, G., Li, Z., Si, J., Xu, K., Cai, Y., Xu, D., & Al-Dhahir, N. (2023). Analysis and optimization of STAR-RIS-assisted proactive eavesdropping with statistical CSI. *IEEE Transactions on Vehicular Technology, 72*, 6850–6855.

Jia, H., Ma, L., & Valaee, S. (2023). STAR-RIS enabled downlink secure NOMA network under imperfect CSI of eavesdroppers. *IEEE Communications Letters, 27*, 802–806.

Jiang, J., Lyu, B., Chen, P., & Yang, Z. (2023). Active STAR-RIS assisted wireless information and power transfer systems. In *Proceedings of the IEEE 98th Vehicular Technology Conference (VTC-Fall)* (pp. 1–7).

Lei, J., Zhang, T., & Liu, Y. (2024). Hybrid NOMA for STAR-RIS enhanced communication. *IEEE Transactions on Vehicular Technology, 73*, 1497–1502.

Li, H., Liu, Y., Mu, X., Chen, Y., Pan, Z., & Eldar, Y. C. (2023a). Near-field beamforming for STAR-RIS networks. arXiv preprint arXiv:2306.14587 .

Li, X., Gao, X., Yang, L., Liu, H., Wang, J., & Rabie, K. M. (2023b). Performance analysis of STAR-RIS-CR-NOMA based consumer IoT networks for resilient industry 5.0. *IEEE Transactions on Consumer Electronics, 70*(1), 1380–1391.

Liu, Y., Wang, Z., Xu, J., Ouyang, C., Mu, X., & Schober, R. (2023a). Near-field communications: A tutorial review. *IEEE Open Journal of the Communications Society, 4*, 1999–2049.

Liu, Y., Zhang, S., Mu, X., Ding, Z., Schober, R., Al-Dhahir, N., Hossain, E., & Shen, X. (2022). Evolution of NOMA toward next generation multiple access (NGMA) for 6G. *IEEE Journal on Selected Areas in Communications, 40*, 1037–1071.

Liu, Z., Li, Z., Wen, M., Gong, Y., & Wu, Y.-C. (2023b). STAR-RIS-aided mobile edge computing: Computation rate maximization with binary amplitude coefficients. *IEEE Transactions on Communications, 71*, 4313–4327.

Mao, Y., You, C., Zhang, J., Huang, K., & Letaief, K. B. (2017). A survey on mobile edge computing: The communication perspective. *IEEE Communications Surveys & Tutorials, 19*, 2322–2358.

Mu, X., Xu, J., Liu, Y., & Hanzo, L. (2024). Reconfigurable intelligent surface-aided near-field communications for 6G: Opportunities and challenges. *IEEE Vehicular Technology Magazine, 19*, 65–74.

Niu, H., Chu, Z., Zhou, F., & Zhu, Z. (2021). Simultaneous transmission and reflection reconfigurable intelligent surface assisted secrecy MISO networks. *IEEE Communications Letters, 25*, 3498–3502.

Ponnimbaduge Perera, T. D., Jayakody, D. N. K., Sharma, S. K., Chatzinotas, S., & Li, J. (2018). Simultaneous wireless information and power transfer (SWIPT): Recent advances and future challenges. *IEEE Communications Surveys & Tutorials, 20*, 264–302.

Qin, X., Song, Z., Hou, T., Yu, W., Wang, J., & Sun, X. (2023). Joint resource allocation and configuration design for STAR-RIS-enhanced wireless-powered MEC. *IEEE Transactions on Communications, 71*, 2381–2395.

Rivetti, S., Demir, O. T., Björnson, E., & Skoglund, M. (2024). Malicious reconfigurable intelligent surfaces: How impactful can destructive beamforming be? *IEEE Wireless Communications Letters*.

Vu, T.-H., Nguyen, T.-V., Pham, Q.-V., Benevides da Costa, D., & Kim, S. (2023). STAR-RIS-enabled short-packet NOMA systems. *IEEE Transactions on Vehicular Technology, 72*, 13764–13769.

Wang, Q., Pang, X., Wu, C., Xu, L., Zhao, N., & Yu, F. R. (2023). Transmit power minimization for STAR-RIS aided FD-NOMA networks. *IEEE Transactions on Vehicular Technology*. https://doi.org/10.1109/TVT.2023.3322297

Wang, T., Fang, F., & Ding, Z. (2023). Joint phase shift and beamforming design in a multi-user MISO STAR-RIS assisted downlink NOMA network. *IEEE Transactions on Vehicular Technology, 72*, 9031–9043.

Wang, W., Ni, W., Tian, H., Yang, Z., Huang, C., & Wong, K.-K. (2022). Safeguarding NOMA networks via reconfigurable dual-functional surface under imperfect CSI. *IEEE Journal of Selected Topics in Signal Processing, 16*, 950–966.

Wei, Z., Li, B., & Guo, W. (2023). Adversarial reconfigurable intelligent surface against physical layer key generation. *IEEE Transactions on Information Forensics and Security, 18*, 2368–2381.

Wen, H., Tota Khel, A. M., & Hamdi, K. A. (2023). Phase shift configuration strategies for unbalanced T&R users in STAR-RIS-aided NOMA. *IEEE Communications Letters, 27*, 3404–3408.

Wu, C., Mu, X., Liu, Y., Gu, X., & Wang, X. (2022). Resource allocation in STAR-RIS-aided networks: OMA and NOMA. *IEEE Transactions on Wireless Communications, 21*, 7653–7667.

Wu, C., You, C., Liu, Y., Han, S., & Di Renzo, M. (2023). Two-timescale design for STAR-RIS aided NOMA systems. *IEEE Transactions on Communications*. https://doi.org/10.1109/TCOMM.2023.3321737

Xie, Z., Liu, Y., Yi, W., Wu, X., & Nallanathan, A. (2023). Physical layer security for STAR-RIS-NOMA: A stochastic geometry approach. *IEEE Transactions on Wireless Communications*.

Xu, J., Mu, X., & Liu, Y. (2024). Exploiting STAR-RISs in near-field communications. *IEEE Transactions on Wireless Communications, 23*, 2181–2196.

Xu, J., Yuan, L., Yang, N., Yang, N., & Guo, Y. (2023a). Performance analysis of STAR-IRS aided NOMA short-packet communications with statistical CSI. *IEEE Transactions on Vehicular Technology, 72*, 12385–12390.

Xu, W., Chen, J., & Yu, X. (2023b). Joint design of power allocation and amplitude coefficients for ergodic rate optimization in STAR-RIS-aided NOMA system. *IEEE Systems Journal, 17*, 5452–5463.

Yang, S., Zhang, J., Xia, W., Gao, H., & Zhu, H. (2022). Joint power and discrete amplitude allocation for STAR-RIS-aided NOMA system. *IEEE Transactions on Vehicular Technology, 71*, 13382–13386.

Yaswanth, J., Katwe, M., Singh, K., Taghizadeh, O., Schmeink, A., & Pan, C. (2023). Joint beamforming design for STAR-RIS-aided MU-MIMO system with SWIPT. In *Proceedings of the IEEE International Conference on Communications (ICC Workshops)* (pp. 574–579).

Yue, X., Xie, J., Liu, Y., Han, Z., Liu, R., & Ding, Z. (2023). Simultaneously transmitting and reflecting reconfigurable intelligent surface assisted NOMA networks. *IEEE Transactions on Wireless Communications, 22*, 189–204.

Zeng, Y., Wu, Q., & Zhang, R. (2019). Accessing from the sky: A tutorial on UAV communications for 5G and beyond. *Proceedings of the IEEE, 107*, 2327–2375.

Zhai, X., Han, G., Cai, Y., Liu, Y., & Hanzo, L. (2023). Simultaneously transmitting and reflecting (STAR) RIS assisted over-the-air computation systems. *IEEE Transactions on Communications, 71*, 1309–1322.

Zhang, C., Yi, W., Liu, Y., Ding, Z., & Song, L. (2022a). STAR-IOS aided NOMA networks: Channel model approximation and performance analysis. *IEEE Transactions on Wireless Communications, 21*, 6861–6876.

Zhang, H., Nie, J., Yu, Y., Xiong, Z., Jiang, W., & Niyato, D. (2023a). Performance analysis for STAR-RIS assisted SWIPT system over rayleigh fading channel. In *Proceedings of the IEEE International Conference on Communications (ICC)* (pp. 1468–1473).

Zhang, H., Shlezinger, N., Guidi, F., Dardari, D., Imani, M. F., & Eldar, Y. C. (2022b). Beam focusing for near-field multiuser MIMO communications. *IEEE Transactions on Wireless Communications, 21*, 7476–7490.

Zhang, N., Liu, Y., Mu, X., Wang, W., & Huang, A. (2023). Queue-aware STAR-RIS assisted NOMA communication systems. *IEEE Transactions on Wireless Communications*. https://doi.org/10.1109/TWC.2023.3322381

Zhang, Q., Wang, Y., Li, H., Hou, S., & Song, Z. (2023b). Resource allocation for energy efficient STAR-RIS aided MEC systems. *IEEE Wireless Communications Letters, 12*, 610–614.

Zhang, S., Zhang, Y., & Di, B. (2024). Near-far field codebook design for IOS-aided multi-user communications. arXiv preprint arXiv:2401.08165 .

Zhang, Z., Wang, Z., Liu, Y., He, B., Lv, L., & Chen, J. (2023c). Security enhancement for coupled phase-shift STAR-RIS networks. *IEEE Transactions on Vehicular Technology, 72*, 8210–8215.

Zhou, F., You, C., & Zhang, R. (2021). Delay-optimal scheduling for IRS-aided mobile edge computing. *IEEE Wireless Communications Letters, 10*, 740–744.

Zhou, T., Xu, K., Hu, G., Xia, X., Xie, W., & Li, C. (2023). Robust beamforming design for STAR-RIS-assisted anti-jamming and secure transmission. *IEEE Transactions on Green Communications and Networking*. https://doi.org/10.1109/TGCN.2023.3329127

Zhu, G., Mu, X., Guo, L., Huang, A., & Xu, S. (2023). Robust resource allocation for STAR-RIS assisted SWIPT systems. *IEEE Transactions on Wireless Communications*. https://doi.org/10.1109/TWC.2023.3327502

Zhu, J., Gao, P., Chen, G., Xiao, P., & Quddus, A. (2023). Index modulation for STAR-RIS assisted NOMA system. *IEEE Communications Letters, 27*, 716–720.

Zuo, J., Liu, Y., Ding, Z., Song, L., & Poor, H. V. (2023). Joint design for simultaneously transmitting and reflecting (STAR) RIS assisted NOMA systems. *IEEE Transactions on Wireless Communications 22*, 611–626.

Appendix A
Chapter 4: Proof of Theorem 4.2

Since the expressions for the outage probabilities for OMA and NOMA in (4.24), (4.29), and (4.30) have similar forms, the respective proofs required for Theorem 4.2 can be given at the same time. According to the system model presented in Sect. 4.3.1, both $|g_m|$ and $|h_m^R|$ follow Rician distributions. Hence, $h_m = \beta^R |g_m||h_m^R|$ is the product of two independent Rician random variables. The PDF of $|h_m|$ can be expressed as follows (Simon, 2002):

$$f_{|h_m|}(x) = \frac{x}{(\beta^R)^2 \beta_h \beta_g} e^{-(K_h+K_g)}$$
$$\cdot \sum_{i=0}^{\infty} \sum_{l=0}^{\infty} \frac{\left(\frac{\alpha_h \sqrt{x}}{2\beta_h}\right)^{2i} \left(\frac{\alpha_g \sqrt{x}}{2\beta_g}\right)^{2l}}{i!l!\Gamma(i+1)\Gamma(l+1)} \left(\frac{\beta_h}{\beta_g}\right)^{\frac{i-l}{2}} K_{i-l}\left(\frac{x}{\beta^R \sqrt{\beta_h \beta_g}}\right), \tag{A.1}$$

where $\alpha_{h/g}^2 = \frac{K_{h/g}\Omega_{h/g}}{K_{h/g}+1}$, $\beta_{h/g} = \frac{\Omega_{h/g}}{2(K_{h/g}+1)}$, $\Gamma(x)$ denotes the Gamma function, and K_n is the modified Bessel function of the second kind. For the asymptotic behavior, we consider the PDF of $|h_m|$ near the origin using a Taylor series expansion:

$$f_{|h_m|}(x) = \frac{e^{-(K_h^R+K_g)}}{(\beta^R)^2 \beta_h \beta_g} \cdot x + o(x), \tag{A.2}$$

where $o(\cdot)$ is the little-o notation and $o(f(x))$ denotes a function which is asymptotically smaller than $f(x)$. The Laplace transform of the asymptotic PDF in (A.2) is given by

$$\mathcal{M}_{|h_m|}(t) = \mathcal{L}\{f_{|h_m|}\} = \frac{1}{(\beta^R)^2 \beta_h \beta_g} e^{-(K_h^R+K_g)} t^{-2} + o(t^{-2}). \tag{A.3}$$

Y. Liu et al., *Simultaneously Transmitting and Reflecting Surfaces for Wireless Communications*, Wireless Networks, https://doi.org/10.1007/978-3-031-76848-4

According to (4.33), the overall channel of user R is the summation of $M+1$ terms. Using the convolution theorem of the Laplace transform, we obtain the Laplace transform of the asymptotic PDF of H^R as follows:

$$\mathcal{M}_{|H^R|}(t) = \mathcal{L}\{f_{|H^R|}\} = \left(\frac{e^{-(K_h+K_g)}}{(\beta^R)^2\beta_h\beta_g}\right)^M \cdot \frac{2(K_d^R+1)}{\Omega_d^R}$$

$$\cdot e^{-K_d^R} t^{-2M-2} + o(t^{-2M-2}). \tag{A.4}$$

Finally, $f_{|H^R|}(x)$ can be obtained by performing the inverse transform term by term. By further substituting $\beta_{h/g}$ with $\frac{\Omega_{h/g}}{K_{h/g}+1}$, we arrive at

$$f_{|H^R|}(x) = \frac{2^{M+1}(K_h^R+1)^M(K_g+1)^M(K_d+1)}{(\beta^R)^{2M}\Omega_h^M\Omega_g^M\Omega_d}$$

$$\cdot e^{-MK_h^R-MK_g-K_d} \cdot \frac{x^{2M+1}}{(2M+1)!}. \tag{A.5}$$

Finally, Theorem 4.2 can be proved by integrating the asymptotic PDF in (A.5) according to (4.29) and (4.24) for NOMA and OMA.

Reference

Simon, M. K. (2002). *Probability distributions involving Gaussian random variables: A handbook for engineers and scientists*. Springer.

Appendix B
Chapter 4: Proof of Lemma 4.1

As shown in Sect. 4.3, for the secondary user for the PS-PSC strategy and both users for the DP-PSC strategy, the STARS-aided channel (h_s^χ) can be expressed as follows:

$$h_s^\chi = |g|\beta^\chi \sum_{m=1}^{M} |h_m^\chi| \exp\{j(\angle h_d^\chi + \delta_m^\chi)\}, \tag{B.1}$$

where δ_m^χ models the phase error of the m-th element for user χ. For both considered PSC strategies, this phase error is bounded, i.e., $|\delta_p^\chi - \delta_q^\chi| \leq \Delta^\chi, \forall p, q$. For the PS-PSC strategy, $\Delta^T = \pi$, and for the DP-PSC strategy, $\Delta^R = \Delta^T = \pi/2$. Hence, as was shown in Danufane et al. (2021, Corollary 2), the PDF of the magnitude of h_s^χ can be approximated by a Rician distribution:

$$f_{|h_s^\chi|}(x) = \frac{x}{\beta^2} e^{-\frac{x^2+\alpha^2}{2\beta^2}} I_0\left(\frac{x\alpha}{\beta}\right), \tag{B.2}$$

where

$$\alpha = M\mathbb{E}[|H_m|]\mathrm{sinc}\left(\frac{\Delta^\chi}{2}\right),\ \beta^2 = \frac{M}{2}\mathbb{E}[|H_m|^2][1 - \mathrm{sinc}(\Delta^\chi)]. \tag{B.3}$$

Thus, for the PS-PSC strategy, the PDF of $|h_s^\chi|$ is obtained by plugging $\Delta^T = \pi$ into (B.2) and (B.3).

Y. Liu et al., *Simultaneously Transmitting and Reflecting Surfaces for Wireless Communications*, Wireless Networks, https://doi.org/10.1007/978-3-031-76848-4

Reference

Danufane, F. H., Renzo, M. D., de Rosny, J., & Tretyakov, S. (2021). On the path-loss of reconfigurable intelligent surfaces: An approach based on green's theorem applied to vector fields. *IEEE Transactions on Communications, 69*, 5573–5592.

Appendix C
Chapter 5: Proof of Theorem 5.1

The relaxed version of problem (5.15) without rank-one constraint (5.8e) is jointly concave with respect to $\left\{\mathbf{W}_k, \mathbf{Q}_k, \boldsymbol{\beta}_k\right\}$ and satisfies Slater's constraint qualification (Boyd & Vandenberghe, 2004). Therefore, strong duality holds, and the Lagrangian function is given by

$$\mathcal{L}=\sum\nolimits_{k\in\{t,r\}} \mu_k \left(\frac{\overline{\gamma}_k}{2}\left\|\mathbf{Q}_k + \mathbf{H}_k \mathbf{W}_{\overline{k}} \mathbf{H}_k^H\right\|_F^2 - \overline{\gamma}_k \mathrm{Tr}\left(\left(\mathbf{H}_k^H \mathbf{H}_k \mathbf{W}_{\overline{k}}^{(n)} \mathbf{H}_k^H \mathbf{H}_k\right)^H \mathbf{W}_{\overline{k}}\right)\right.$$
$$\left. + \sum\nolimits_{k\in\{t,r\}} \mathrm{Tr}\left(\mathbf{W}_k\right) + \frac{1}{2}\left\|\mathbf{Q}_k - \mathbf{H}_k \mathbf{W}_k \mathbf{H}_k^H\right\|_F^2 - \mathrm{Tr}\left(\left(\mathbf{H}_k^H \mathbf{H}_k \mathbf{W}_k^{(n)} \mathbf{H}_k^H \mathbf{H}_k\right)^H \mathbf{W}_k\right)\right)$$
$$- \sum\nolimits_{k\in\{t,r\}} \mathrm{Tr}\left(\mathbf{Y}_k \mathbf{W}_k\right) + \tau, \tag{C.1}$$

where τ is the collection of all terms that do not depend on $\{\mathbf{W}_k\}$, and μ_k and $\mathbf{Y}_k$ are the Lagrange multipliers associated with constraints (5.15b) and (5.8f), respectively. Based on the Karush-Kuhn-Tucker (KKT) conditions with respect to $\mathbf{W}_k$, the structure of the optimal $\mathbf{W}_k^*$ can be characterized as follows:

$$\mu_k^* \geq 0, \mathbf{Y}_k^* \succeq \mathbf{0}, \tag{C.2a}$$
$$\mathbf{Y}_k^* \mathbf{W}_k^* = \mathbf{0}, \tag{C.2b}$$
$$\nabla_{\mathbf{W}_k^*} \mathcal{L} = \mathbf{0}, \tag{C.2c}$$

where μ_k^* and $\mathbf{Y}_k^*$ denote the optimal Lagrange multipliers and $\nabla_{\mathbf{W}_k^*} \mathcal{L}$ is the gradient of $\mathcal{L}$ with respect to $\mathbf{W}_k^*$. Then, $\nabla_{\mathbf{W}_k^*} \mathcal{L} = 0$ can be further expressed as

$$\mathbf{Y}_k^* = \mathbf{I}_N - \boldsymbol{\Omega}_k^*, \tag{C.3}$$

Y. Liu et al., *Simultaneously Transmitting and Reflecting Surfaces for Wireless Communications*, Wireless Networks, https://doi.org/10.1007/978-3-031-76848-4

where $\mathbf{\Omega}_k^*$ is given by

$$\mathbf{\Omega}_k^* = \mu_k \left(\mathbf{H}_k \mathbf{Q}_k \mathbf{H}_k^H - \mathbf{H}_k \mathbf{H}_k \mathbf{W}_k \mathbf{H}_k^H \mathbf{H}_k^H + \mathbf{H}_k^H \mathbf{H}_k \mathbf{W}_k^{(n)} \mathbf{H}_k^H \mathbf{H}_k \right)$$
$$-\mu_{\overline{k}} \overline{\gamma}_{\overline{k}} \left(\mathbf{H}_{\overline{k}} \mathbf{Q}_k \mathbf{H}_{\overline{k}}^H + \mathbf{H}_{\overline{k}} \mathbf{H}_{\overline{k}} \mathbf{W}_k \mathbf{H}_{\overline{k}}^H \mathbf{H}_{\overline{k}}^H - \mathbf{H}_{\overline{k}}^H \mathbf{H}_{\overline{k}} \mathbf{W}_k^{(n)} \mathbf{H}_{\overline{k}}^H \mathbf{H}_{\overline{k}} \right). \tag{C.4}$$

By exploiting the results in (Xu et al., 2020, Appendix A), it can be proved that Rank $\left(\mathbf{Y}_k^*\right) = N-1$. Furthermore, (C.2b) implies that Rank $\left(\mathbf{Y}_k^*\right)$+Rank $\left(\mathbf{W}_k^*\right) \leq N$. Therefore, Rank $\left(\mathbf{W}_k^*\right) \leq 1$ has to hold. Thus, due to the QoS constraint in (5.15b), Rank $\left(\mathbf{W}_k^*\right) = 1$ for the optimal solution. This completes the proof.

References

Boyd, S., & Vandenberghe, L. (2004). *Convex optimization*. Cambridge University Press.

Xu, D., Yu, X., Sun, Y., Ng, D. W. K., & Schober, R. (2020). Resource allocation for IRS-assisted full-duplex cognitive radio systems. *IEEE Transactions on Communications, 68*, 7376–7394.

Appendix D
Chapter 5: Proof of Proposition 5.2

For any given $\tilde{\boldsymbol{\beta}}_t$ and $\tilde{\boldsymbol{\beta}}_r$, problem (5.42) can be decomposed into a series of independent optimization problems for each pair of $(\tilde{\psi}_{t,n}, \tilde{\psi}_{r,n})$, which leads to

$$\min_{\tilde{\psi}_{t,n}, \tilde{\psi}_{r,n}} \quad \mathrm{Re}(\tilde{\vartheta}_{t,n}^* \tilde{\psi}_{t,n}) + \mathrm{Re}(\tilde{\vartheta}_{r,n}^* \tilde{\psi}_{r,n}) \tag{D.1a}$$

$$\text{s.t.} \quad \cos(\tilde{\phi}_{t,n} - \tilde{\phi}_{r,n}) = 0, \tag{D.1b}$$

$$|\tilde{\psi}_{t,n}| = 1, |\tilde{\psi}_{r,n}| = 1. \tag{D.1c}$$

Note that the coupled phase-shift constraint (D.1b) can be rewritten as $|\tilde{\phi}_{t,n} - \tilde{\phi}_{r,n}| = \frac{\pi}{2}$ or $\frac{3\pi}{2}$, which is equivalent to

$$\tilde{\psi}_{r,n} = j\tilde{\psi}_{t,n} \text{ or } \tilde{\psi}_{r,n} = -j\tilde{\psi}_{t,n}. \tag{D.2}$$

Substituting the above constraint into the objective function, problem (D.1) can be further simplified as $\min_{|\tilde{\psi}_{t,n}|=1} \mathrm{Re}\left((\tilde{\vartheta}_{t,n}^* \pm j\tilde{\vartheta}_{r,n}^*)\tilde{\psi}_{t,n}\right)$, the optimal solution to which can be readily obtained as follows:

$$\tilde{\psi}_{t,n} = e^{j\left(\pi - \angle(\tilde{\vartheta}_{t,n}^* \pm j\tilde{\vartheta}_{r,n}^*)\right)}. \tag{D.3}$$

Comparing the objective values for the above two solutions, the optimal solution to problem (D.1) can be obtained, which completes the proof.

Y. Liu et al., *Simultaneously Transmitting and Reflecting Surfaces for Wireless Communications*, Wireless Networks, https://doi.org/10.1007/978-3-031-76848-4

Appendix E
Chapter 5: Proof of Proposition 5.3

For any given $\tilde{\boldsymbol{\psi}}_t$ and $\tilde{\boldsymbol{\psi}}_r$, problem (5.42) can be decomposed into a series of independent optimization problems for each pair of $(\tilde{\beta}_{t,n}, \tilde{\beta}_{r,n})$, which leads to

$$\min_{\tilde{\beta}_{t,n},\tilde{\beta}_{r,n}} \quad \mathrm{Re}(\breve{\vartheta}^*_{t,n}\tilde{\beta}_{t,n}) + \mathrm{Re}(\breve{\vartheta}^*_{r,n}\tilde{\beta}_{r,n}) \tag{E.1a}$$

$$\text{s.t.} \quad \tilde{\beta}^2_{t,n} + \tilde{\beta}^2_{r,n} = 1, 0 \le \tilde{\beta}_{t,n}, \tilde{\beta}_{r,n} \le 1. \tag{E.1b}$$

Since $\tilde{\beta}_{i,n}$ is real-valued, we simplify the objective function as follows: $a_n\tilde{\beta}_{t,n} + b_n\tilde{\beta}_{r,n}$, where $a_n = |\breve{\vartheta}^*_{t,n}|\cos(\angle\breve{\vartheta}^*_{t,n})$ and $b_n = |\breve{\vartheta}^*_{r,n}|\cos(\angle\breve{\vartheta}^*_{r,n})$. To solve the simplified problem, polar coordinates are used, where we set $\tilde{\beta}_{t,n} = \sin\omega_n$ and $\tilde{\beta}_{r,n} = \cos\omega_n$ with $\omega_n \in [0, \frac{1}{2}\pi]$. Hence, constraint (E.1b) is automatically satisfied. Based on this transformation, the objective function can be rewritten as

$$\begin{aligned} & a_n \sin\omega_n + b_n \cos\omega_n \\ =& \sqrt{a_n^2 + b_n^2}\,(\cos\xi_n \sin\omega_n + \sin\xi_n \cos\omega_n) \\ =& \sqrt{a_n^2 + b_n^2}\,\sin(\omega_n + \xi_n), \end{aligned} \tag{E.2}$$

where $\cos\xi_n = \frac{a_n}{\sqrt{a_n^2+b_n^2}}$ and $\sin\xi_n = \frac{b_n}{\sqrt{a_n^2+b_n^2}}$. Then, the optimal solution in (5.46) for minimizing $\sin(\omega_n + \xi_n)$ with respect to $\omega_n \in [0, \frac{1}{2}\pi]$ can be readily obtained, which completes the proof.

Y. Liu et al., *Simultaneously Transmitting and Reflecting Surfaces for Wireless Communications*, Wireless Networks, https://doi.org/10.1007/978-3-031-76848-4

Appendix F
Chapter 5: Proof of Proposition 5.4

In this appendix, we show that the MFCQ condition (Shi et al., 2020, Appendix C) is satisfied by problem (5.52). Note that for problem (5.52), optimizing $\tilde{\boldsymbol{\theta}}_i$ is equivalent to optimizing its amplitudes $\tilde{\boldsymbol{\beta}}_i = [\tilde{\beta}_{i,1}, \ldots, \tilde{\beta}_{i,N}]$ and phase shifts $\tilde{\boldsymbol{\phi}}_i = [\tilde{\phi}_1, \ldots, \tilde{\phi}_N]$. Thus, all equality constants in problem (5.52) can be written as

$$\boldsymbol{\mu}_1 = \tilde{\boldsymbol{\beta}}_t \circ \tilde{\boldsymbol{\beta}}_t + \tilde{\boldsymbol{\beta}}_r \circ \tilde{\boldsymbol{\beta}}_r - \mathbf{1} = \mathbf{0}, \tag{F.1a}$$

$$\boldsymbol{\mu}_2 = \cos(\tilde{\boldsymbol{\phi}}_t - \tilde{\boldsymbol{\phi}}_r) = \mathbf{0}, \tag{F.1b}$$

$$\boldsymbol{\mu}_3 = \tilde{\boldsymbol{\beta}}_t \circ e^{j\tilde{\boldsymbol{\phi}}_t} - \boldsymbol{\theta}_t = \mathbf{0}, \tag{F.1c}$$

$$\boldsymbol{\mu}_4 = \tilde{\boldsymbol{\beta}}_r \circ e^{j\tilde{\boldsymbol{\phi}}_r} - \boldsymbol{\theta}_r = \mathbf{0}, \tag{F.1d}$$

where $\circ$ denotes the element-wise product, $\mathbf{1}$ is the all-ones vector, and $\mathbf{0}$ is the all-zeros vector. First, we show that the gradients of $\{\boldsymbol{\mu}_i\}_{i=1}^4$ with respect to the vector of variables $\boldsymbol{\omega} = \left[\tilde{\boldsymbol{\beta}}_t^T, \tilde{\boldsymbol{\beta}}_r^T, \tilde{\boldsymbol{\phi}}_t^T, \tilde{\boldsymbol{\phi}}_r^T, \boldsymbol{\theta}_t^T, \boldsymbol{\theta}_t^T\right]^T$ are linearly independent. It can be observed that since only $\boldsymbol{\mu}_4$ contains variable $\boldsymbol{\theta}_r$, the gradients of $\{\boldsymbol{\mu}_i\}_{i=1}^4$ are linearly independent if and only if those of $\{\boldsymbol{\mu}_i\}_{i=1}^3$ are linearly independent. Similarly, since only $\boldsymbol{\mu}_3$ contains variable $\boldsymbol{\theta}_t$, the gradients of $\{\boldsymbol{\mu}_i\}_{i=1}^3$ are linearly independent if and only if those of $\{\boldsymbol{\mu}_i\}_{i=1}^2$ are linearly independent. Furthermore, the gradients of $\{\boldsymbol{\mu}_i\}_{i=1}^2$ are linearly independent since $\boldsymbol{\mu}_2$ does not contain $\tilde{\boldsymbol{\beta}}_t$ and $\tilde{\boldsymbol{\beta}}_r$ but $\boldsymbol{\mu}_1$ does. As a consequence, the gradients of $\{\boldsymbol{\mu}_i\}_{i=1}^4$ must be linearly independent. Then, according to the definition of MFCQ (Shi et al., 2020, Appendix C), we are left to show that there exists a matrix $\mathbf{D_W}$ and a vector $\mathbf{d}_\omega$, such that

$$\mathrm{Re}\{\mathrm{tr}(\mathbf{W}\mathbf{D}_{\mathbf{W}}^H)\} < 0, \tag{F.2a}$$

$$\nabla\boldsymbol{\mu}_i \mathbf{d}_\omega = \mathbf{0}, \forall i \in \{1, 2, 3, 4\}, \tag{F.2b}$$

Y. Liu et al., *Simultaneously Transmitting and Reflecting Surfaces for Wireless Communications*, Wireless Networks, https://doi.org/10.1007/978-3-031-76848-4

where $\nabla \boldsymbol{\mu}_i$ denotes the Jacobian matrix of $\boldsymbol{\mu}_i$ with respect to $\boldsymbol{\omega}$. It can be readily proved that the above equations are satisfied by setting $\mathbf{D_W} = -\mathbf{W}$ and $\mathbf{d}_\omega = \mathbf{0}$, which completes the proof.

Reference

Shi, Q., Hong, M., Fu, X., & Chang, T.-H. (2020). Penalty dual decomposition method for nonsmooth nonconvex optimization—part II: Applications. *IEEE Transactions on Signal Processing, 68*, 4242–4257.

Appendix G
Chapter 7: Maximum Likelihood Estimate of DOAs

In this appendix, the MLE of DOAs is derived. To facilitate the estimation at the sensors, we assume that the transmit signal $\mathbf{X}$ and the channel $\mathbf{G}$ are known at the sensors, which can be obtained by the wired control link between the BS and the STARS. According to (7.13), the vectorized signal over a coherent time block of length L at the sensor can be rewritten as

$$\mathbf{y}_s = \alpha \boldsymbol{\delta}(\phi_h, \phi_v) + \mathbf{n}_s, \tag{G.1}$$

where $\boldsymbol{\delta}(\phi_h, \phi_v) = \mathrm{vec}(\mathbf{b}(\phi_h, \phi_v)\mathbf{a}^T(\phi_h, \phi_v)\boldsymbol{\Theta}_r \mathbf{G}\mathbf{X})$. It can be observed that $\mathbf{y}_s$ is a Gaussian vector with mean $\alpha\boldsymbol{\delta}(\phi_h, \phi_v)$ and variance $\sigma_s \mathbf{I}_{N_S L}$. Given parameters $\boldsymbol{\xi}$, the likelihood function of $\mathbf{y}_s$ is

$$f_{\mathbf{y}_s}(\mathbf{y}_s; \boldsymbol{\xi}) = \frac{1}{\sqrt{(\pi\sigma_s)^{N_S L}}} \exp\left(-\frac{1}{\sigma^2}\|\mathbf{y}_s - \alpha\boldsymbol{\delta}(\phi_h, \phi_v)\|^2\right). \tag{G.2}$$

Thus, the MLE of $\boldsymbol{\xi}$ is given by

$$\hat{\boldsymbol{\xi}} = \arg\max_{\boldsymbol{\xi}} f_{\mathbf{y}_s}(\mathbf{y}_s; \boldsymbol{\xi}) = \arg\min_{\boldsymbol{\xi}} \|\mathbf{y}_s - \alpha\boldsymbol{\delta}(\phi_h, \phi_v)\|^2. \tag{G.3}$$

According to (G.3), for any given ϕ_h and ϕ_v, α can be estimated as

$$\hat{\alpha} = \arg\min_{\alpha} \|\mathbf{y}_s - \alpha\boldsymbol{\delta}(\phi_h, \phi_v)\|^2 = \frac{\boldsymbol{\delta}^H(\phi_h, \phi_v)}{\|\boldsymbol{\delta}(\phi_h, \phi_v)\|^2}\mathbf{y}_s. \tag{G.4}$$

With $\hat{\alpha}$ at hand, we have

$$\|\mathbf{y}_s - \hat{\alpha}\boldsymbol{\delta}(\phi_h, \phi_v)\|^2 = \|\mathbf{y}_s\|^2 - \frac{|\boldsymbol{\delta}^H(\phi_h, \phi_v)\mathbf{y}_s|^2}{\|\boldsymbol{\delta}(\phi_h, \phi_v)\|^2}. \tag{G.5}$$

Y. Liu et al., *Simultaneously Transmitting and Reflecting Surfaces for Wireless Communications*, Wireless Networks, https://doi.org/10.1007/978-3-031-76848-4

Thus, the MLE of ϕ_h and ϕ_v is given by

$$(\hat{\phi}_h, \hat{\phi}_v) = \arg\max_{\phi_h, \phi_v} \frac{|\boldsymbol{\delta}^H(\phi_h, \phi_v)\mathbf{y}_s|^2}{\|\boldsymbol{\delta}(\phi_h, \phi_v)\|^2}, \tag{G.6}$$

which can be obtained by exhaustively searching ϕ_h and ϕ_v on the fine grids of $[0, \pi]$ and $[-\frac{\pi}{2}, \frac{\pi}{2}]$, respectively.

Appendix H
Chapter 7: Derivation of the Fisher Information Matrices

It can be readily observed that $\mathbf{y}_s$ is a Gaussian observation with the distribution $\mathbf{y}_s \sim \mathcal{CN}(\mathbf{u}, \mathbf{R}_n)$, where $\mathbf{R}_n = \sigma_s^2 \mathbf{I}_{N_s L}$ represents the covariance matrix of $\mathbf{n}_s$. Then, following the same path in Kay (1993, Appendix 3C), the element at the ℓ-th row and the p-th column of $\mathbf{J}_{\boldsymbol{\xi}}$ can be calculated by

$$\begin{aligned}[\mathbf{J}_{\boldsymbol{\xi}}]_{\ell,p} =& 2\mathrm{Re}\left\{\frac{\partial \mathbf{u}^H}{\partial \xi_\ell}\mathbf{R}_n^{-1}\frac{\partial \mathbf{u}}{\partial \xi_p}\right\} + \mathrm{tr}\left(\mathbf{R}_n^{-1}\frac{\partial \mathbf{R}_n}{\partial \xi_\ell}\mathbf{R}_n^{-1}\frac{\partial \mathbf{R}_n}{\partial \xi_p}\right) \\ =& \frac{2}{\sigma_s^2}\mathrm{Re}\left\{\frac{\partial \mathbf{u}^H}{\partial \xi_\ell}\frac{\partial \mathbf{u}}{\partial \xi_p}\right\},\end{aligned} \tag{H.1}$$

where ξ_ℓ denotes the ℓ-th element of $\boldsymbol{\xi}$. Thus, we first derive the derivative of $\mathbf{u}$ with respect to the unknown parameters. By defining $\mathbf{B} = \mathbf{b}(\phi_h, \phi_v)\mathbf{a}^T(\phi_h, \phi_v)$, we have

$$\frac{\partial \mathbf{u}}{\partial \boldsymbol{\phi}} = [\alpha \mathrm{vec}(\dot{\mathbf{B}}_{\phi_h}\boldsymbol{\Theta}_r \mathbf{G}\mathbf{X}), \alpha \mathrm{vec}(\dot{\mathbf{B}}_{\phi_v}\boldsymbol{\Theta}_r \mathbf{G}\mathbf{X})], \tag{H.2}$$

$$\frac{\partial \mathbf{u}}{\partial \tilde{\boldsymbol{\alpha}}} = \mathrm{vec}(\mathbf{B}\boldsymbol{\Theta}_r \mathbf{G}\mathbf{X})[1, j], \tag{H.3}$$

where

$$\begin{aligned}\dot{\mathbf{B}}_{\phi_h} =& \frac{\partial \mathbf{B}}{\partial \phi_h} = \frac{\partial \mathbf{b}}{\partial \phi_h}\mathbf{a}^T + \mathbf{b}\frac{\partial \mathbf{a}^T}{\partial \phi_h} \\ =& j\frac{2\pi}{\lambda_c}\sin\phi_h \cos\phi_v \left(\mathrm{diag}(\bar{\mathbf{r}}_X)\mathbf{b}\mathbf{a}^T + \mathbf{b}\mathbf{a}^T \mathrm{diag}(\mathbf{r}_X)\right).\end{aligned} \tag{H.4}$$

$$\dot{\mathbf{B}}_{\phi_v} = \frac{\partial \mathbf{B}}{\partial \phi_v} = \frac{\partial \mathbf{b}}{\partial \phi_v}\mathbf{a}^T + \mathbf{b}\frac{\partial \mathbf{a}^T}{\partial \phi_v}$$

Y. Liu et al., *Simultaneously Transmitting and Reflecting Surfaces for Wireless Communications*, Wireless Networks, https://doi.org/10.1007/978-3-031-76848-4

$$
\begin{aligned}
&= j\frac{2\pi}{\lambda_c}\cos\phi_h \sin\phi_v\left(\operatorname{diag}(\bar{\mathbf{r}}_X)\mathbf{b}\mathbf{a}^T + \mathbf{b}\mathbf{a}^T\operatorname{diag}(\mathbf{r}_X)\right) \\
&\quad - j\frac{2\pi}{\lambda_c}\cos\phi_v\mathbf{b}\mathbf{a}^T\operatorname{diag}(\mathbf{r}_Z).
\end{aligned} \tag{H.5}
$$

In the above formulas, we drop ϕ_h and ϕ_v in $\mathbf{b}(\phi_h, \phi_v)$ and $\mathbf{a}(\phi_h, \phi_v)$ for notational convenience. Then, the matrix $\mathbf{J}_{\boldsymbol{\phi}\boldsymbol{\phi}}$ can be further partitioned as

$$
\mathbf{J}_{\boldsymbol{\phi}\boldsymbol{\phi}} = \begin{bmatrix} J_{\phi_h\phi_h} & J_{\phi_h\phi_v} \\ J_{\phi_h\phi_v} & J_{\phi_v\phi_v} \end{bmatrix}. \tag{H.6}
$$

According to (H.1), the entries $J_{\phi_l\phi_p}, \forall l, p \in \{h, v\}$, of the matrix $\mathbf{J}_{\boldsymbol{\phi}\boldsymbol{\phi}}$ can be calculated as follows:

$$
\begin{aligned}
J_{\phi_l\phi_p} &= \frac{2}{\sigma_s^2}\operatorname{Re}\left\{\alpha^*\operatorname{vec}(\dot{\mathbf{B}}_{\phi_l}\boldsymbol{\Theta}_r\mathbf{G}\mathbf{X})^H\alpha\operatorname{vec}(\dot{\mathbf{B}}_{\phi_p}\boldsymbol{\Theta}_r\mathbf{G}\mathbf{X})\right\} \\
&= \frac{2|\alpha|^2 L}{\sigma_s^2}\operatorname{Re}\left\{\operatorname{tr}(\dot{\mathbf{B}}_{\phi_p}\boldsymbol{\Theta}_r\mathbf{G}\mathbf{R}_x\mathbf{G}^H\boldsymbol{\Theta}_r^H\dot{\mathbf{B}}_{\phi_l}^H)\right\}.
\end{aligned} \tag{H.7}
$$

Next, the matrices $\mathbf{J}_{\boldsymbol{\phi}\tilde{\boldsymbol{\alpha}}}$ and $\mathbf{J}_{\tilde{\boldsymbol{\alpha}}\tilde{\boldsymbol{\alpha}}}$ are derived as follows:

$$
\begin{aligned}
\mathbf{J}_{\boldsymbol{\phi}\tilde{\boldsymbol{\alpha}}} &= \frac{2}{\sigma_s^2}\operatorname{Re}\left(\begin{bmatrix}\alpha^*\operatorname{vec}(\dot{\mathbf{B}}_{\phi_h}\boldsymbol{\Theta}_r\mathbf{G}\mathbf{X})^H \\ \alpha^*\operatorname{vec}(\dot{\mathbf{B}}_{\phi_v}\boldsymbol{\Theta}_r\mathbf{G}\mathbf{X})^H\end{bmatrix}\operatorname{vec}(\mathbf{B}\boldsymbol{\Theta}_r\mathbf{G}\mathbf{X})[1, j]\right) \\
&= \frac{2L}{\sigma_s^2}\operatorname{Re}\left(\begin{bmatrix}\alpha^*\operatorname{tr}\left(\mathbf{B}\boldsymbol{\Theta}_r\mathbf{G}\mathbf{R}_x\mathbf{G}^H\boldsymbol{\Theta}_r^H\dot{\mathbf{B}}_{\phi_h}^H\right) \\ \alpha^*\operatorname{tr}\left(\mathbf{B}\boldsymbol{\Theta}_r\mathbf{G}\mathbf{R}_x\mathbf{G}^H\boldsymbol{\Theta}_r^H\dot{\mathbf{B}}_{\phi_v}^H\right)\end{bmatrix}[1, j]\right),
\end{aligned} \tag{H.8}
$$

$$
\begin{aligned}
\mathbf{J}_{\tilde{\boldsymbol{\alpha}}\tilde{\boldsymbol{\alpha}}} &= \frac{2}{\sigma_s^2}\operatorname{Re}\left((\operatorname{vec}(\mathbf{B}\boldsymbol{\Theta}_r\mathbf{G}\mathbf{X})[1, j])^H\operatorname{vec}(\mathbf{B}\boldsymbol{\Theta}_r\mathbf{G}\mathbf{X})[1, j]\right) \\
&= \frac{2}{\sigma_s^2}\operatorname{Re}\left([1, j]^H[1, j]\operatorname{vec}(\mathbf{B}\boldsymbol{\Theta}_r\mathbf{G}\mathbf{X})^H\operatorname{vec}(\mathbf{B}\boldsymbol{\Theta}_r\mathbf{G}\mathbf{X})\right) \\
&= \frac{2L}{\sigma_s^2}\mathbf{I}_2\operatorname{tr}\left(\mathbf{B}\boldsymbol{\Theta}_r\mathbf{G}\mathbf{R}_x\mathbf{G}^H\boldsymbol{\Theta}_r^H\mathbf{B}^H\right).
\end{aligned} \tag{H.9}
$$

Reference

KAY, S. M. (1993). *Fundamentals of statistical signal processing: Estimation theory*. Prentice-Hall.

Appendix I
Chapter 7: Proof of Proposition 7.3

For any given $\tilde{\boldsymbol{\beta}}_t$ and $\tilde{\boldsymbol{\beta}}_r$, the optimization problem with respect to $\tilde{\mathbf{q}}_t$ and $\tilde{\mathbf{q}}_r$ is given by

$$\min_{\tilde{\mathbf{q}}_t, \tilde{\mathbf{q}}_r} \quad \mathrm{Re}(\tilde{\boldsymbol{\upsilon}}_t^H \tilde{\mathbf{q}}_t) + \mathrm{Re}(\tilde{\boldsymbol{\upsilon}}_r^H \tilde{\mathbf{q}}_r) \tag{I.1a}$$

$$\text{s.t.} \quad [\tilde{\mathbf{q}}_r]_n = j[\tilde{\mathbf{q}}_t]_n \text{ or } [\tilde{\mathbf{q}}_r]_n = -j[\tilde{\mathbf{q}}_t]_n, \forall n, \tag{I.1b}$$

$$|[\tilde{\mathbf{q}}_t]_n| = 1, |[\tilde{\mathbf{q}}_r]_n| = 1, \forall n, \tag{I.1c}$$

where constraint (I.1b) is transformed from the coupled phase-shift constraint (7.54c). It can be observed this problem is a separable optimization problem. In other words, each pair of $(\tilde{q}_{t,n}, \tilde{q}_{r,n})$ can be optimized individually, and the related optimization problem is given by

$$\min_{\tilde{q}_{t,n}, \tilde{q}_{r,n}} \quad \mathrm{Re}(\tilde{\upsilon}_{t,n}^* \tilde{q}_{t,n}) + \mathrm{Re}(\tilde{\upsilon}_{r,n}^* \tilde{q}_{r,n}) \tag{I.2a}$$

$$\text{s.t.} \quad \tilde{q}_{r,n} = j\tilde{q}_{t,n} \text{ or } \tilde{q}_{r,n} = -j\tilde{q}_{t,n}, \tag{I.2b}$$

$$|\tilde{q}_{t,n}| = 1, |\tilde{q}_{r,n}| = 1. \tag{I.2c}$$

Substituting the constraint (I.2b) into the objective function, the above problem can be further simplified as $\min_{|\tilde{q}_{t,n}|=1} \mathrm{Re}\left((\tilde{\upsilon}_{t,n}^* \pm j\tilde{\upsilon}_{r,n}^*)\tilde{q}_{t,n}\right)$, where the factor $(\tilde{\upsilon}_{t,n}^* + j\tilde{\upsilon}_{r,n}^*)$ is for the case $\tilde{q}_{r,n} = j\tilde{q}_{t,n}$ and the factor $(\tilde{\upsilon}_{t,n}^* - j\tilde{\upsilon}_{r,n}^*)$ is for the case $\tilde{q}_{r,n} = -j\tilde{q}_{t,n}$. It is clear that the optimal $\tilde{q}_{t,n}$ is given by

$$\tilde{q}_{t,n} = e^{j\left(\pi - \angle(\tilde{\upsilon}_{t,n}^* \pm j\tilde{\upsilon}_{r,n}^*)\right)}. \tag{I.3}$$

By matching the optimal $\tilde{q}_{t,n}$ with the cases of $\tilde{q}_{r,n} = j\tilde{q}_{t,n}$ and $\tilde{q}_{r,n} = -j\tilde{q}_{t,n}$, the solutions in (7.58) can be obtained, which completes the proof.

Y. Liu et al., *Simultaneously Transmitting and Reflecting Surfaces for Wireless Communications*, Wireless Networks, https://doi.org/10.1007/978-3-031-76848-4

Appendix J
Chapter 7: Proof of Proposition 7.4

For any given $\tilde{\mathbf{q}}_t$ and $\tilde{\mathbf{q}}_r$, the optimization problem with respect to $\tilde{\boldsymbol{\beta}}_t$ and $\tilde{\boldsymbol{\beta}}_r$ is given by

$$\min_{\tilde{\boldsymbol{\beta}}_t, \tilde{\boldsymbol{\beta}}_r} \quad \mathrm{Re}(\breve{\boldsymbol{\upsilon}}_t^H \tilde{\boldsymbol{\beta}}_t) + \mathrm{Re}(\breve{\boldsymbol{\upsilon}}_r^H \tilde{\boldsymbol{\beta}}_r) \tag{J.1a}$$

$$\text{s.t.} \quad \tilde{\beta}_{t,n}^2 + \tilde{\beta}_{r,n}^2 = 1, 0 \leq \tilde{\beta}_{t,n}, \tilde{\beta}_{r,n} \leq 1, \forall n, \tag{J.1b}$$

which is also a separable problem. The separated problem related to $\tilde{\beta}_{t,n}$ and $\tilde{\beta}_{r,n}$ is given by

$$\min_{\tilde{\beta}_{t,n}, \tilde{\beta}_{r,n}} \quad g_n = \mathrm{Re}(\breve{\upsilon}_{t,n}^* \tilde{\beta}_{t,n}) + \mathrm{Re}(\breve{\upsilon}_{r,n}^* \tilde{\beta}_{r,n}) \tag{J.2a}$$

$$\text{s.t.} \quad \tilde{\beta}_{t,n}^2 + \tilde{\beta}_{r,n}^2 = 1, 0 \leq \tilde{\beta}_{t,n}, \tilde{\beta}_{r,n} \leq 1, \tag{J.2b}$$

where $\breve{\upsilon}_{i,n}, \forall i \in \{t, r\}$, denotes the n-th entry of $\breve{\boldsymbol{\upsilon}}_i$. Since $\tilde{\beta}_{i,n}, \forall i \in \{t, r\}$, is real-valued, the objective function can be further simplified as $a_n \tilde{\beta}_{t,n} + b_n \tilde{\beta}_{r,n}$, where $a_n = |\breve{\upsilon}_{t,n}^*| \cos(\angle \breve{\upsilon}_{t,n}^*)$ and $b_n = |\breve{\upsilon}_{r,n}^*| \cos(\angle \breve{\upsilon}_{r,n}^*)$. In this case, problem (J.2) is essentially to find the minimum value of the real-valued function $a_n \tilde{\beta}_{t,n} + b_n \tilde{\beta}_{r,n}$ on the unit circle $\tilde{\beta}_{t,n}^2 + \tilde{\beta}_{r,n}^2 = 1$ in the first quadrant. To solve it, we transform it into the polar coordinate system by setting $\tilde{\beta}_{t,n} = \sin \omega_n$ and $\tilde{\beta}_{r,n} = \cos \omega_n$ with $\omega_n \in [0, \frac{1}{2}\pi]$. Then, the objective function can be rewritten as

$$\begin{aligned} g_n &= a_n \sin \omega_n + b_n \cos \omega_n \\ &\overset{(a)}{=} \sqrt{a_n^2 + b_n^2} \, (\cos \psi_n \sin \omega_n + \sin \psi_n \cos \omega_n) \\ &= \sqrt{a_n^2 + b_n^2} \sin(\omega_n + \psi_n), \end{aligned} \tag{J.3}$$

Y. Liu et al., *Simultaneously Transmitting and Reflecting Surfaces for Wireless Communications*, Wireless Networks, https://doi.org/10.1007/978-3-031-76848-4

where the equality (a) is achieved by defining $\cos\psi_n = \frac{a_n}{\sqrt{a_n^2+b_n^2}}$ and $\sin\psi_n = \frac{b_n}{\sqrt{a_n^2+b_n^2}}$. As a consequence, the problem is to find the minimum value of $\sin(\omega_n + \psi_n)$ with respect to ω_n in the interval $[0, \frac{1}{2}\pi]$. Thus, the optimal ω_n in (5.54) can be readily obtained. Based on this, the optimal $\tilde{\beta}_{t,n}$ and $\tilde{\beta}_{r,n}$ can also be obtained, which completes the proof.

The manufacturer's authorised representative in the EU is Springer
Nature Customer Service Centre GmbH, Europaplatz 3, 69115 Heidelberg,
Germany. If you have any concerns regarding our products, please
contact ProductSafety@springernature.com

Printed and bound by CPI Group (UK) Ltd, Croydon, CR0 4YY
01/10/2026
02225966-0006